Predictive Technology in Social Media

Editors

Cristina Fernández-Rovira

Department of Communication, University of Vic-Central University of Catalonia, Barcelona, Spain

Santiago Giraldo-Luque

Department of Journalism and Communication

Autonomous University of Barcelona, Barcelona, Spain

CRC Press

Taylor & Francis Group

Boca Raton London New York

CRC Press is an imprint of the Taylor & Francis Group, an **informa** business

A SCIENCE PUBLISHERS BOOK

First edition published 2022
by CRC Press
6000 Broken Sound Parkway NW, Suite 300, Boca Raton, FL 33487-2742

and by CRC Press
2 Park Square, Milton Park, Abingdon, Oxon, OX14 4RN

CRC Press is an imprint of Taylor & Francis Group, LLC

ISBN: 978-1-032-10340-2 (hbk)
ISBN: 978-1-032-10345-7 (pbk)
ISBN: 978-1-003-21487-8 (ebk)

DOI: 10.1201/9781003214878

Typeset in Times New Roman
by Innovative Processors

Acknowledgements

The editors would especially like to thank Vijay Primlani and the entire CRC PRESS team for the opportunity to publish this book. Likewise, the editors would also like to thank the authors of the chapters who have contributed with their interesting contents.

Preface

The months it takes in the editorial process to prepare a book determine its content. The world can change suddenly and make the text obsolete. It may also be that the book predicts what will happen just in the time that pass while it is being prepared for publication. The battle to capture the attention and decisions of users through technology, as well as the dispute over the possibility of predicting and shaping the future, has only just begun. Facebook Inc. has changed its name and, at the same time, has made the most blatant bid for the domination of social, economic and cultural relations. The presentation of the Metaverse in October 2021, just as this book was being prepared for publication, opens the door wide to the content of its pages, which now become much more relevant.

Predictive Technology in Social Media takes a critical look at the main dangers and potential of the use of technology to define, control and guide the future life of societies. Written by academics from different universities around the world, which allows for a global approach to the problem of technological domination, the text becomes an essential tool for understanding how technology controls and can predict the most intimate behaviour of individuals. The work explores the problems of the predictive capacity of technology in relation to its algorithmic definition, its potential to attract and control human attention, its complex scheme of advertising management, the type of audio-visual consumption carried out by people, its use within digital culture, its application to politics and social mobilisation, its media impact and its ethical structuring. All these problems have gone beyond the technical perspective and the seductive interface, and have ended up affecting the type of relations between people, and between people and machines.

The book approaches these aspects of the universe of technological prediction from a mainly qualitative, explanatory point of view. It is a warning reading to avoid, among other things, the destruction of free will, the disappearance of identities (individual and collective), the collapse of will or personal values and the disappearance of sensorial, human socialisation. It is an invitation to think about a social relationship reconverted into perfect avatars-profiles constructed with millions of sensors that now flood the intimacy and privacy of spaces and bodies.

The text reflects, from a multidisciplinary perspective, on a very near future in which humans cannot act as humans under their own freedom of choice. At the beginning of the 21st Century, the book is a warning about the disappearance of the Charter of Rights that built the liberal and welfare society. But it is also a call to propose, to think and to build, from ethics and human values, a necessary and urgent Charter of Digital Rights, linked to digital wellbeing and the protection of people from the dominant platforms that have completely taken over the contemporary digital ecosystem.

Cristina Fernández-Rovira
Santiago Giraldo-Luque

Introduction: From Delphi to Zuckerberg – Conquering the Future

> *"So they put forth their hands to the good cheer lying ready before them. But when they had put from them the desire of food and drink, the Muse moved the minstrel to sing of the glorious deeds of warriors, from that lay the fame whereof had then reached broad heaven, [75] even the quarrel of Odysseus and Achilles, son of Peleus, how once they strove with furious words at a rich feast of the gods, and Agamemnon, king of men, was glad at heart that the best of the Achaeans were quarrelling; for thus Phoebus Apollo, in giving his response, had told him that it should be, [80] in sacred Pytho, when he passed over the threshold of stone to enquire of the oracle. For then the beginning of woe was rolling upon Trojans and Danaans through the will of great Zeus". Odyssey, Book 8.*

In the *Odyssey*, almost 3000 years ago, Agamemnon seeks the advice of the Oracle at Delphi before setting sail for Troy. This was the usual practice taken before any important decision had to be made in the life of the Ancient Greeks. During the classical period, Delphi had become the most popular oracle in the Hellenic world. It was so famous that its halo of divination reached as far as Egypt and Asia Minor. Travellers came from far and wide to consult the future. Kings, politicians, cities, businessmen and warriors submitted to the wisdom of the temple of Delphi. For the pharaohs, the parallel between Apollo, God of the temple, and Amun Ra, their own solar deity, was obvious.

The sanctuary of Delphi is so ancient that its origins are lost in mythology. Legend has it that it was consecrated to Gaia, goddess of the earth, and her daughter Themis, until Apollo arrived, deceived by the nymph Delphusa. She had assured him that the place was ideal for founding an oracle, without warning him that a monstrous serpent lived there. Apollo defeated the dragon by the Castalia spring and left its body to rot (popular etymology relates the term rot to the Greek verb "pythomai", hence the relation between the death and decomposition of the serpent and the word python and, consequently, the word pythoness). After punishing the nymph , the god went in search for the first priests for his temple.

To do this, according to mythology, Apollo took the form of a dolphin, jumped aboard a ship from Crete, which he guided to the coast and offered the sailors to a chance to enter his service in the temple, as priests. From then on, the area was given the name Delphi (Greek for dolphin). Apollo would also be known by the nickname Delphicus. It is possible that the story of the boat and the dolphin has some historical basis, as the island of Crete is believed to be the place from which the cult of Apollo spread throughout Greece (Echeverría Arístegui 2019).

The sanctuary also attracted intellectuals: Plutarch was a priest at Delphi, Pythagoras trained a priestess, and Socrates ironically claimed that the oracle had named him the wisest among men precisely because he recognised his ignorance (Echeverría Arístegui 2019). It is difficult to know why Delphi was more successful than other religious centres, such as the oracles of Zeus at Olympia or Dodona (the latter also consulted in the *Odyssey*).

According to legend, King Croesus of Lydia once tested the most famous oracles by sending them all the same question. Only Delphi got it right. It was also said that the sanctuary stood on the exact spot where two eagles, sent by Zeus from opposite ends of the earth, had crossed their flight, right over the slope of Mount Parnassus. At that point the king of Olympus deposited the Omphalos, the navel of the world (Martínez 2015).

Delphi was therefore considered the centre of the known world. Two oval stones (the omphalos), one inside the temple of Apollo and one outside, reminded visitors of its history. Beyond the myth, Delphi was the heart of classical Greece, the most influential and well-informed destination of its time. Every day news came from distant regions, carried by those who came to consult the priestesses.

Although pilgrims flocked to the sanctuary in search of guidance on decisions to be taken, the essential function of the oracle was not to predict the future, but to provide divine sanction to the political decisions of the cities. It ratified laws and even constitutions, approved the founding of new cities and colonies, advised on warlike enterprises, or censured them (Movellán Luis 2017). And although Delphi did not intervene directly in the politics of the cities, its oracles could be used as a political weapon if necessary.

Although the Oracle of Delphi originally offered consultations only once a month and never in winter, its popularity meant that its fortune-telling services multiplied. "The auguries became more and more frequent, and at the height of their popularity, two pythonesses attended to visitors simultaneously, while a third waited her turn" (Echeverría Arístegui 2019).

The interest in knowing the future and the large congregations of citizens who flocked to the oracle catapulted the power and wealth of Delphi. At the same time, the temple became an important focus of political influence. Different territories (first Chrysa, and later Lacedemonia, Athens, Delphi and Phocis) engaged in sacred wars that ended with the neutrality of the oracle. History also meant that conflicts with the Persians, the Galatians and among the Hellenic peoples played a part in the temple's loss of credibility: "it inevitably lost its aura of neutrality. The Greeks ceased to trust its omens blindly, at least in political matters (...) and the oracle went into a slow decline" (Echeverría Arístegui 2019).

The fascination with knowing or predicting the future has always had a magical and inexplicable ingredient for human beings. Palm readers, crystal balls, prophets, seers, witches, pythonesses, and other mythological or religious beings have tried to claim the gift of predicting what might happen. In the same way, divination of the thoughts of others, mind-reading, or the discovery of the desires of others are faculties that, from fiction, have given heroes or imaginary characters supernatural powers that give them dominion over the will of other people.

Scientifically, predictive functions (weather, rainfall, natural disasters, diseases, pandemics, election results or a country's GDP growth) determine people's daily decisions by reading data, facts, and available information. Both the tarot (irrationally) and the weatherman (with scientific and rational precision) influence the behaviour of individuals. They are agents of prediction.

On a historical level, algorithmic predictive capability is at the root of the beginning of computing. Alan Turing formulated one of the main questions about prediction in his essay: *Can a Machine Think?* (1974). Although the technology was still in its infancy and the capacity to produce, collect and analyse data was small, Turing predicted that technological science would advance so far that it was only a matter of time before technology would be able to make accurate predictions, based on information and data, about everyday human operations. His *Enigma* machine, capable of deciphering the encrypted messages of the Nazi forces in World War II, was a pioneer in reading, analysing, and processing information to predict the future, based on data. Tragically, Turing, who can be considered one of the main pioneers of the conceptualisation and development of the algorithm, was prosecuted for homosexuality in 1952, a fact that ended his brilliant career and, two years later, his life.

Enigma's predictive power gave the Allied forces a significant advantage and, according to Jack Copeland, professor at the University of Canterbury, its work saved millions of lives by significantly reducing the duration of the war between two to four years (Copeland 2012).

The fascination that hides the power to know the future or to know how to predict, delineate or lead it, has also marked the course of technology since the consolidation of the second wave of the internet, in which it is the users who produce, every second, information and data that are recorded by supercomputers. After the failure of the first internet, which occurred with the dot.com bubble at the end of the millennium, the new collaborative web or 2.0 O'Reilly and Battelle (2009) were able to identify new economic strategies for internet companies. The fundamental framework of the new strategy was found in leveraging user-generated data (or content) to establish additional values on that previously unanalysed, free, and highly valuable product.

Google and Amazon, the only major companies to benefit from the turn of the millennium, were the pioneers in approaching the prediction domain and gradually perfected the model after Amazon patented the system called "Collaborative Recommendations Using Item-to-Item Similarity Mappings" (Linden et al.1998). The era of user-informed prediction dominance had dawned.

Based on the Amazon model (and Google's initially functional Page Rank system), the large, consolidated technology companies (Google, Amazon, Facebook, Apple, and Microsoft, to which we can add some Chinese companies and other minor unicorns such as Netflix and Twitter) are engaged in a battle for control of users' screen time. This screen time within their platforms (fuelled by free content produced by users) means that they can have control of the advertising market and the monopolistic establishment of the pricing system under a single dominant and absolutist mandate. But for now, the new framework of competition is for control of the user's next phase. For the ability to predict what the user wants, thinks, and intends to be.

In a further step, the dispute of the third decade of the 21st century is focused on determining which technological company is more capable of delineating and manipulating, through the highly subtle intrusion of users' everyday on-screen spaces, what the user wants, thinks or pretends to be.

In this path of prediction and control, social media networks are the protagonists. Their social validation systems, freely accepted by the masses of users, have become modern-day Delphic oracles. The idea has been installed in society that, if you want to be somebody, you have to go to social networks and "sell" your image, your idea, your content, your news, your political ideology, within the platform. They, the platforms, have set themselves up (with the legitimacy of the world's population) as the experts of what the whole world can see, can know, and can think. They censor what does not suit them when they want. They lead the way in social discussion and, at the same time, play a systemic role that keeps dissent, transformation, and social revolution under control, as this too must be sold on the networks.

The concentrated power of Delphi, as a social validator of decisions (individual, social, economic, collective) has become technologically sophisticated in the 21st century in social media. The power and data concentration capacity of big technology is infinite. It is fed every second by millions and millions of bytes of data that are processed by algorithms. Every second, a decision is predicted or suggested. The system has become so sophisticated that Alexa, with a sleek, modern design, like an object of desire, answers all the user's questions, making life easier. In the process emptying human life.

According to James Williams (2021) the Android smartphone operating system sends more than eleven billion notifications a day to its more than one billion users. Distraction, free reflection, and autonomous decision-making are becoming endangered species. The constant calls for attention, which are designed under perfectly structured fields of psychological emotionality, are persuasive invitations for the user to perform the actions that the platforms expect them to perform because they are perfectly designed for them. They in turn predict them.

In their predictive capacity, social media platforms win the battle for control, while the subject disappears and is lost in the programmed automatisms of algorithms. In their systemic perspective, algorithms act by reducing the complexity of the environment by internally acquiring more data that feeds the capacity for prediction, the domination over the user who becomes an automaton and dependent on the technological response. Without it, they do not know where to go, what places to visit, what music to listen to, which partner to choose or what they want to be when they grow up. It is certainly a sign of the times that many young people actually declare that they want to be Youtubers, Tiktokers or Instagramers.

The self-definition of future will, or individual projection assimilated under the name of a technological platform denotes the capacity of dominant companies to predict and install a will alien to the individual (with its values, its ideologies, its mechanisms). This goes so far as to convert it into a must be, in the definition of "wanting what we want to want" (Williams 2021), of the new and not so new generations.

Fortunately, as it happened in Delphi, in recent years and from different sectors, critical voices have begun to be raised against the dynamics of the control systems of social media and other large technology companies. The problems of social manipulation, lack of transparency, deregulation of their functions, their fiscal manoeuvres to pay less taxes, their policies for handling user data, political polarisation, and hundreds of manifestations of physical and psychological effects, especially on children and young people, are becoming more and more frequent and increasingly loud.

As in Delphi, the aura of neutrality is no longer associated with any social media network. Citizens are beginning to position themselves as important critics of its effects and the trust placed in its promise of social recognition is beginning to waver. It would not be bad news if the temples of narcissism, individualism, and the expression of xenophobic, sexist, racist and extremist ideas were to enter, like Delphi, into a slow decline.

Can our behaviour on social media predict a future purchase? Can the clicks we make on social networks predict which political party we will vote for? Can the information we share on our wall foretell the next series we will consume? Can the likes we give on Instagram and Facebook announce the time we will spend on social media platforms in the next hour?

The answer is no longer science fiction. It defines the ability of major social media companies, such as Facebook and Twitter, to deliver specialised advertising services to highly targeted audience segments controlled by the billions of devices that pervade our daily lives.

At the same time, it highlights a more relevant problem: can social media networks use all the behavioural information to guide, suggest or impose a certain behaviour or thinking? All indications are that it is possible. They already predict behaviours.

The book *Predictive Technology in Social Media* is composed of nine essays and studies that reflect on the power of predictive social media technology in culture, entertainment, marketing and economics, politics, and the ethics of free will. The text has the challenge of analysing from a humanistic, diverse, and critical perspective the predictive possibilities of social media networks, as well as the risks they entail for cultural and ideological plurality, everyday consumption, the monopolistic concentration of economy and attention, and representative democracy.

The book wants to be an invitation to think, as citizens, about the unbridled power that we have ceded to social media networks. A new reflection to alert us to the greatest concentration of communicative power ever seen in the history of humanity. To this end, the book is divided into three main parts.

The first part deals with the general and preliminary debates that shape the discussions about algorithmic culture and on the functioning of attention, the raw material through which user data is extracted. The first chapter of the book, written by Diego García Ramírez, Professor of Journalism and Public Opinion Programme at the Universidad Del Rosario (Bogotá, Colombia) and Dune Valle Jiménez, Professor at the School of Communication Sciences at the Universidad Sergio Arboleda (Bogotá, Colombia), proposes a critical approach to the algorithms and their implications in people's daily lives. The chapter, entitled "Algorithmic Culture: Limits and Notes for the Discussion", highlights that algorithms increasingly play a more important role in the way in which human beings relate to others and experience the world, as well as in the sources from which they build the meanings that guide behaviours in both online and offline environments. García Ramírez and Valle Jiménez present an invitation to understand algorithms beyond their technical characteristics and begin to understand their cultural implications.

The second chapter of the book, entitled "The Functioning of Attention as a Behavioural Prediction Mechanism in Social Media" has been developed by Pedro Nicolás Aldana Afanador, Professor of the Social Communication Programme at the Autonomous University of Bucaramanga (Colombia). The text explores the concept of attention from different scientific perspectives and approaches the functioning of attentional networks from the studies of neuropsychology and neurophysiology. Likewise, with the aim of defining the importance of attention captured by new social networks technologies for the establishment of predictions about user behaviour, the text describes the main characteristics of attention and how they are used by social networks precisely to capture, concentrate, and secure the attention of users.

The second part of the book focuses on explaining how different platforms manage user data to predict and guide users' consumption patterns and behaviour on certain products or platforms. Chapter three of the book, "Predictive Analytics in Digital Advertising: Are Platforms Really Interested in Accuracy?" is written by Òscar Coromina, Senior Lecturer in Media Technology at Malmö University (Sweden), and examines the predictive features that digital platforms make available to marketing and advertising professionals to create and manage digital advertising campaigns. Using a series of interviews, the text explains how digital marketing practitioners provide an account of the state of the art of predictive analytics: in which processes intervene, how reliable they are and to what extent they have been adopted by professionals. The chapter also identifies some limitations on the tools and techniques that digital platforms make available to marketing professionals to prepare their digital advertising campaigns.

Emiliano Lucas Iglesia Albores, Ph.D student in Communication and Journalism at the Autonomous University of Barcelona (Spain), proposes in the fourth chapter a theoretical and empirical approach to describe how Netflix utilizes user information with the intention of predicting future consumption. The chapter, entitled "Consumption Prediction on Netflix: Audience Tracking Analysis Based on the Recommendation Algorithm in Times of Pandemic", reviews the concepts of algorithm, database and artificial intelligence and then describes the results of a six-month consumption monitoring study of twelve Netflix users. The developed study demonstrates that the greater the consumption in time and productions on the platform, the greater the power of Netflix to generate more accurate consumption suggestions that are accepted by users. The text also stresses that the platform's power to predict future consumption, fuelled by the pandemic effect, determines a scenario of control over human decision-making that extends, on other internet platforms, to all life experiences.

The fifth chapter of the book, entitled "Social Complex Networks Analysis as Predictors of Users' Behaviour in the Digital Society", focuses on a methodology to visualise the data produced by web analytics as networks and nodes, which allow it to analyse the complexity of the relationships between social actors and identify their actions, strategies and innovations oriented towards the promotion and offer in social media networks, which can guide the design of marketing strategies. The chapter is written by Laura Isabel Rojas de Francisco, Juan Carlos Monroy Osorio and Santiago Rodríguez Cadavid, professors and researchers of the Marketing Department of the Universidad EAFIT (Medellín, Colombia). It exposes the methodological procedures to explain a form of analysis concentrated on graph based representation of texts in which the content shared on social networking sites provides the transcript.

The last part of the book explores the ethical, philosophical and political implications of the prediction exercised by social media networks today. This third part of the book begins with the chapter "Predicting Government Attention in Social Media: A First Step for Understanding Political *astroturf* in Interest Representation", written by Camilo Cristancho Mantilla, Professor of the Department of Political Science and Public Administration, at University of Barcelona (Spain). The study describes the extent in which the government and presidential agenda can be predicted by the issues given the attention of interest groups using Twitter data from active interest group organisations and the official users from the government cabinet between March 2018 and March 2021 in Spain. The chapter develops an aggregate analysis for all types of interest group organisations and shows that the presidential agenda can be better predicted than the government agenda by the activity of interest group organisations. Results of the study provide relevant insights for the regulation of artificial intelligence in the activity of political actors.

The seventh chapter explores the impact of prediction in the construction of contemporary social mobilisation. The text "Social Media as a Framework for Predicting and Controlling Social Protest in the 21st Century" has been elaborated by Santiago Giraldo Luque, Professor of the Department of Journalism and Communication Studies, at Autonomous University of Barcelona (Spain). The chapter proposes a critical theoretical reading of the use of social media as the main space for communication and articulation of the objectives of contemporary social mobilisations. Based on a theoretical analysis, the text explores how social media has undermined the opportunities for social change promoted by the dispersed and individualised mobilisation that is channelled through social media platforms. The study also criticises the absence of strategies for social mobilisation, the centrality of actions on the handling, management, and obsession for trending topics on social media, and the limited temporal permanence of social mobilisations due to the absence of a collective identity.

"Reviving Topological Thinking in the Post Media Condition" is the title of the eighth chapter of the book. The text, written by Justin Michael Battin, Lecturer of the Professional Communications Department, at RMIT University (Ho Chi Minh City, Vietnam), argues that the realisation promised by a post-media condition must account for the notion that the internet, and social media specifically, paradigmatically perpetuates the collapse of meaningful distinctions. Battin states that all information is flexible, and all media objects are interchangeable. Yet, while the promise of a post-media condition is not without hurdles, it remains pursuable through topological thinking, made possible by the articulation of place and place-making practices and a reinvigoration of a care-oriented perspective that centres the disclosive character of human being.

Finally, the chapter "Ethical Insights for the Social Media Age" discusses at how the practice of predicting behaviour (and the gradual loss of free will) has been introduced in large technological platforms, mainly in social media networks. At the same time, it addresses the ethical perspective implied by the loss of individual and collective decision-making capacity (increasingly externalised in the interfaces) and, finally, it takes a closer look at the new social contract based on a Charter of Digital Rights, a proposal that is beginning to be considered by different national and international public institutions, as well as by some civil society organisations. The text, written by Cristina Fernández-Rovira, Professor of the Department of Communication, at University of Vic-Central University of Catalonia (Vic, Spain), approaches the main ethical debates that must necessarily be introduced in the interpretation of technology as an inherent layer of today's society.

The editors of the book are grateful to all the participating authors, who take a holistic look at the problem of prediction in social networks and take an approach that integrates different countries and scientific disciplines. All of them take a critical look at the power of social media networks as predictive tools in different fields and warn of the concentration of a large part of users' personal information in the hands of a few people.

The essays and studies that make up the book open questions for research and rational reflection which, when guided by the universities, should shed light and provide answers of understanding to one of the main problems of today's media and sociological world. There is still time to propose actions and establish debates so that individual will does not succumb to the magic of prediction and algorithmic behavioural control.

Cristina Fernández-Rovira and Santiago Giraldo-Luque

August, 2021

References

Copeland, J. 2012. Alan Turing: The codebreaker who saved 'millions of lives'. *BBC News*, June 19, 2021. https://www.bbc.com/news/technology-18419691

Echeverría Arístegui, A. 2019. This is how the Delphi oracle worked. *La Vanguardia*, October 17, 2019. https://www.lavanguardia.com/historiayvida/historia-antigua/20191017/47900462955/delfos-gracia-clasica-oraculo.html

Linden, G.D., J.A. Jacobi and E.A. Benson. 1998. *Collaborative Recommendations Using Item-to-Item Similarity Mappings*, US Patent 6,266,649, to Amazon.com, Patent and Trademark Office, 2001 (filed 1998).

Martínez, O. 2015. *Odisea. Índice Onomástico*. Gredos Library.

Movellán Luis, M. 2017. Delphi, the oracle of the god Apollo. *Historia. National Geographic*, June 13, 2017. https://historia.nationalgeographic.com.es/a/delfos-oraculo-dios-apolo_7276

O'Reilly, T. and J. Battelle. 2009. *Web Squared: Web 2.0 Five Years On*. O'Reilly Media, Inc/Web 2.0 Summit.

Turing, A.M. 1974. *Can a Machine Think*? University of Valencia.

Willliams, J. 2021. *Clicks Against Humanity. Freedom and Resistance in the Age of Technological Distraction.* Gatopardo Ensayo.

Contents

Part III: Ethical and Political Implications of Prediction

Part I

General Discussions on Prediction-Oriented Algorithms and Attention

CHAPTER

1

Algorithmic Culture: Limits and Notes for the Discussion[1]

Diego García Ramírez[1]* and Dune Valle Jiménez[2]

[1] Universidad del Rosario, Bogotá, Colombia

[2] Universidad Sergio Arboleda, Bogotá, Colombia

Introduction

Over the last few decades, the discussions around the advances associated with the internet and social media have gained ground within the field of reflection on technology and society. Even though some authors drew attention throughout the 20th century to the need to think about the consequences that unlimited techno-scientific development could have on society (Heidegger 2003, Winner 1980), in the early years of the 21st century these discussions have become more visible due to the fact that Information and Communication Technologies (ICT) permeate all areas of contemporary life.

Usually the reflections around ICTs and their impact swing between exaltation, glorification and celebration, on the one hand; and the contempt, disapproval and revulsion on the other. Namely, between techno-optimists and techno-pessimists, between technophiles and technophobes (Breton 2011, Broussard 2018, Sfez 2005). The former are amazed by the possibilities of technology to solve all the problems facing humanity and the advance towards more democratic, egalitarian, creative, collaborative, intelligent, tolerant and prosperous societies. Meanwhile, the latter see technological innovations as a threat to human beings, who will be

[1] This chapter develops some ideas presented by the authors in an article entitled "Los impactos de la ideología técnica y la cultura algorítmica en la sociedad: una aproximación crítica", published in Revista de Estudios Sociales. See: https://revistas.uniandes.edu.co/doi/full/10.7440/res71.2020.02

*Corresponding author: diegoalo.garcia@urosario.edu.co

replaced and governed by machines, ending not only their freedom and autonomy, but also rendering them useless and unnecessary.

It seems then that in order to intervene in this discussion, one would have to declare to be either a lover or enemy of technology since even halfway positions are frowned upon:

> *"If the authors declare themselves enemies of technology, they are reactionary, retrograde, archaic sclerotic, unable to accept changes [...]. But if, on the contrary, the authors defend technical culture, they turn out to be boastful, naive, unconscious, without a true culture [...] if the author tries to weigh things up and maintain honest moderation, of good law, he lacks ideas, he doesn't know how to choose, he's a softie" (Sfez 2005).*

Phillippe Breton (2011), in a similar sense, expresses that the division between technophobes and technophiles is a fallacy that minimizes and simplifies the debates around technological developments.

Although it seems contradictory, technophiles and technophobes agree on one point: that the best or the worst of the future depends solely and exclusively on technology, hence both are technological determinists. This determinism prevents the understanding of the political, economic and cultural impacts of technology from complexity.

In that context, this chapter proposes a discussion and a critical approach to certain technological developments, specifically, to those related to algorithms and their implications on people's daily lives. It does not seek to indicate if algorithms are good or bad, but to instead highlight that they increasingly play a more important role in the way in which human beings relate to each other and experience the world, in the manner and sources from which we build the senses and meanings that guide our behaviours in both online and offline environments. Thus, it shows how algorithms affect and shape our culture.

The approach to what we will call algorithmic culture will be carried out from a phenomenological and transdisciplinary perspective. "From a phenomenological perspective, approaching algorithms is about being attentive to the ways in which social actors develop more or less reflexive relationships to the systems they are using and how those encounters, in turn, shape online experience" (Bucher 2018). It is transdisciplinary because it takes ideas from some 20th century thinkers of the technique, as well as contributions from philosophy, anthropology, communication and sociology that have addressed the contemporary relationship between technology and society.

At this point, it is important to clarify that the creation and emergence of a new culture is not advocated here, as the promoters of cyber-culture once prophesised (Lévy 2007). The central argument is that contemporary culture continues to be influenced by power relations, inequalities and inequities that are reinforced and/ or legitimized through algorithms.

Algorithms: Computing Machines of the 21st Century

Since the Enlightenment and modernity, the desire of humans to calculate, control and predict the world has become more intense (Mayer-Schönberger and Kenneth 2013, Han, 2014, Crosby 1998). According to the philosopher Yuk Hui, "our way of speaking about progress since the 18th century is marked by the desire to measure, calculate, and dominate" (2020). This was described by Martin Heidegger as "calculative thinking" (2003), thus, part of the developments of the last two centuries in science and technology have been oriented to this purpose, in the hope that a calculable world would be more controllable and, at the same time, better, more efficient, and more productive (Rosa 2020).

Techno-science of the 20th century concentrated itself on developing machines, theories and instruments that would help quantify the world, not only to understand and control it, but also to solve all its problems (Morozov 2015). That was one of the cybernetic goals of the mathematician Nobert Wiener in the middle of the last century (Rid 2016); hereafter Heidegger has seen in cybernetics the most advanced state of calculative thinking.

After World War II, when in addition to cybernetics, the field of computing and Artificial Intelligence was developed, algorithms became part of the scientific language of the time (Berlinski 2000). Despite this, according to some scholars, the term algorithm dates back to the 9th century associated with the Persian mathematician Abdullah Muhammad ibn Musa Al-Khwarizmi (Steiner 2012). However, its use in the field of mathematics did not spread until at least approximately the 17th century, and in the field of computing from the second half of the 20th century.

For these reasons, the starting point must be to strengthen our definition of an algorithm. Put simply, it is like a series of instructions and concrete and finite steps to obtain a particular result (Bucher 2018, Domingos 2015, Finn 2018, O'Neil 2018, Steiner 2012). "An algorithm is an effective procedure, a way of getting something done in a finite number of discrete steps. Classical mathematics is, in part, the study of certain algorithms" (Berlinski, 2000). Algorithms did not appear with informational languages, but it was instead thanks to the development of computer science that their use spread and popularized to occupy the space that they currently have in the contemporary world.

The instructions computational algorithms follow are limited and finite, and are oriented to achieve a concrete result, to offer a solution to the many possibilities that can be obtained. Although an algorithm is a series of instructions, not all algorithms are the same nor do they produce the same results. It all depends on the tasks for which they were programmed and the data they process to fulfil it. According to Dominique Cardon (2016, 2018), there are popularity, authority, reputation and prediction algorithms.

This is the most common and widespread definition, even outside the field of mathematics and computer science; therefore, it is also pertinent to talk about algorithmic culture, because here we are interested in algorithms not only for what they are, but also for what they do: "Social scientists and humanities scholars are not primarily concerned with the technical details of algorithms or their underlying systems but, rather, with the meanings and implications that algorithmic systems may have" (Bucher 2018).

In that sense, algorithms should not be understood only as mathematical artefacts, but rather something cultural. "An algorithm is a perfectly well-defined mathematical object; but it is as well a human artefact, and so an expression of human needs" (Berlinski 2000). No matter how much they are fomented and promoted as objective and neutral tools that only seek to optimize certain processes and for decision-making, algorithms must be understood as cultural artefacts, as producers and reproducers of culture (Gillespie 2016), even if they go unnoticed and seem invisible, algorithms increasingly intervene in the lives of people and societies at large:

> *"Algorithms are in every nook and cranny of civilization. They are woven into the fabric of everyday life. They're not just in your cell phone or your laptop but also in your car, your house, your appliances, and your toys. Your bank is a gigantic tangle of algorithms, with humans turning the knobs here and there. Algorithms schedule flights and then, the airplanes. Algorithms run factories, trade and route goods, cash the proceeds, and keep records" (Domingos 2015).*

The fact that today algorithms play a part in practically all aspects of contemporary life makes them an object of cultural interest, therefore, an object of reflection for the social and human sciences (Manovich 2017).

Despite the fact that algorithms intervene in more and more aspects of everyday life, few know how they function and how they act; however, "even if we do not know what they are and that they are invisible to us, our daily life is increasingly related and conditioned by algorithms" (García and Valle 2020). But aside from the fact that they increasingly impact on culture, algorithms must be understood as cultural artefacts because they are elaborated by people (programmers) who cannot abstract themselves from the contexts and interests under which they elaborate and program them, algorithms carry the print of their creators (Berlinski 2000). For this reason, various female authors have shown that algorithms can become sexist, classist and racist (O'Neil 2018, Eubanks 2021, Noble 2018, Broussard 2018).

Consequently, algorithms reinforce and reproduce historical and hegemonic cultural values such as those associated with gender, social class, and skin colour, among others. "Race, sex, sexual orientation, religion, ability, and other categories have not disappeared online but have become more complicated" (DeNardis 2020). Thus, it's important to approach algorithms as cultural objects, since to look at them simply as useful mathematical processes for the optimization of certain processes, make the consequences of their applications invisible: "But the point is not whether some people benefit. It's that so many suffer. These models, powered by algorithms, slam doors in the face of millions of people, often for the flimsiest of reasons, and offer no appeal. They're unfair" (O´Neil 2018).

A central aspect for the operation and programming of algorithms is the data. With the advancement of the internet and the appearance of applications and platforms (Van Dijck et al. 2018), different sectors and social and individual activities have been digitized, which has impacted not only the way in which they are developed, but the interests behind them. Parallel to this digitization process, a new resource and source of income emerged that is generating changes in almost all sectors of the economy and social life: data. "From representing a peripheral aspect of businesses, data increasingly became a central resource. In the early years of the century it was hardly clear, however, that data would become the raw material to jumpstart a major shift in capitalism" (Srnicek 2018).

The production, capture, storage and analysis of data is known as big data; and for more than a decade in economic, political and academic circles people have been referring to them as the revolution of the 21st century (Cardon 2018, Mayer-Schonberger and Cukier 2013, Stephens-Davidowitz 2019). This increase in the ability to produce, capture, and process data has renewed the "calculative thinking" Heidegger spoke of; thus, finally, thanks to digital technologies, society would have the necessary tools and information not only to understand reality and human problems, but also to solve and correct them.

Thanks to the intermediation of technology, today each users' movement in digital environments can be tracked, registered and converted into data. This process is known as datafication (Couldry 2020, Sadowski 2019, Van Dijck 2014). The datafication the ICTs allow and enhance are part of the dream of Western society to measure and quantify all human and natural phenomena.

Even though the datafication is a process enhanced by digitization, it was not created by it; the desire to measure, calculate, quantify, record, store and process data preceded digitization, "digitization turbocharges datafication. But it is not a substitute. The act of digitization by itself does not datafy" (Mayer-Schonberger and Cukier 2013).

Many activities that are datafied today had escaped the measurement process to which the world has been subjected in recent centuries, this happened for technical reasons or because their commercial value was not yet recognized; however, nowadays no activity or area of individual and social life escapes the datafication process. Today these are datafied: relationships, travels, content consumption, physical activity, and hours of sleep, eating routines, and any other activity mediated by ICTs. In this way, "(meta) data is presented as" raw material "that can be analyzed and processed into predictive algorithms about future human behaviour – valuable assets in the industry" (Van Dijck 2014). People's reality and every aspect of human lives are reduced to their datafiable, calculable and quantifiable character; the world has become nothing more than a collection and accumulation of data.

Although the origin of this form of interpreting reality can be recognized at the dawn of Modernity, now thanks to the developments in technology and the internet, this way of reading, interpreting and understanding the world has acquired greater notoriety and recognition through big data and its corresponding -ism (Han 2014, Lohr 2015, Van Dijck 2014). In other words, dataism renews the belief and trust in the technologies that collect data, in the algorithms that process it, and in automated decision-making through artificial intelligence (AI).

Digitization facilitated the ability to record, store and analyse large amounts of data, as well as to produce data that previously escaped technical possibilities. Dataism is the belief and confidence that any phenomenon and problem can be explained and solved through data. Datafication is the process of converting all human activity mediated by technologies into data.

Data is the *raison d'être* of dataism and datafication; this is why Byung Chul Han compares the contemporary exaltation for big data, with the enthusiasm for statistics in the 18th century. "Then, statistics inspired renewed trust in a higher power for people as they confronted the contingencies of the world" (Han 2014). Big data is based on the data that users produce, but despite this, the ownership of such data rests with the technology companies and not with those who produce it, which has generated debates about the legal implications and the use of those data (Zuboff 2020). In short, the incidence of algorithms in all areas of private and public life has led to user data being used to decide on fundamental aspects of people's lives.

Algorithmization of the Culture

Culture is one of the most complex and elusive concepts in the social and human sciences. Even within anthropology that with over 150 years devoted to its study since the field was defined by Edward B. Taylor, there is no single and homogenous definition that adapts to all studies, periods, and schools of thought (Harris 2001). On the contrary, each theoretical current works with one or more definitions that can sometimes be contradictory.

Already by 1963, the anthropologists Alfred Kroeber and Clyde Kluckhon identified 164 definitions of culture only in the field of American anthropology, which evidenced the expansion of the concept, but also the disagreements around it (Kuper 2001).

Despite the multiplicity of meanings, the symbolic perspective of culture whose main representative is Clifford Geertz, is one of the most widespread and accepted, both in anthropology and in other social sciences (Kottack 2011). For this author, "denotes an historically transmitted pattern of meanings embodied in symbols, a system of inherited conceptions expressed in symbolic forms by means of which men communicate, perpetuate, and develop their knowledge about and attitudes toward life" (Geertz 2003).

The symbolic definition of culture holds that shared meanings are those that allow communication, interaction, and the construction of ideas and images about life and the world. For this reason, as will be pointed out later, today it is the algorithms of social media applications and other services that condition a good part of the meanings from which we relate to other people and with reality. We will call this algorithmic culture.

However, algorithmic culture will not be understood as a new culture as some enthusiasts predicted at the time, believing the ICTS and the web had the possibility of creating new worlds and cultures. With the advances of the internet and the web at the beginning of the last decade of the 20th century, numerous authors and scientists predicted that technologies would allow us, not only to improve the world and societies, but also to create a new place of coexistence and development: the cyberspace (Johnson 2012). This would be a different place to the physical space that human beings inhabited until now[2].

Cyberspace would be a place without borders, states or governments. There would be no hierarchies, fewer inequalities, problems, and conflicts.

> *"As a mythological place, cyberspace was perceived as an infinite and timeless environment, a boundless frontier in which human beings could live in a constant condition of equality, thanks to the horizontal distribution of knowledge and to the creation of the so-called global village. Within this boundless and timeless space, collective freedom and communitarian partnership were the shared values at the core of a new societal organization" (Bory 2020).*

In other words, it would be a better world. The maximum expression of the imaginations and hopes around this new space was "A Declaration of the Independence of Cyberspace" made by John Perry Barlow in Davos, Switzerland, during the 1996 World Economic Forum[3]. The promises and hopes around cyberspace marked the speeches on the progress and development of the new century. Within that idyllic and fantastic place, another type of culture would also be born, cyberculture, which, in the words of the French philosopher Pierre Lévy (2007):

> *"With cyberculture the aspiration to build a social bond is expressed, which would not be based neither on territorial belongings, nor in institutional relations, nor in power relations, but in the reunion around common centres of interest, in playing, in the fact of sharing knowledge, in cooperative learning, in open processes of collaboration."*

This type of culture was not only prophesied as another culture, but as a better one, in which there would be neither hierarchies nor inequalities, everything would

---

[2] The idea of cyberspace takes up the foundations and promises of cybernetics, hence the prefix "cyber".

[3] See "A Declaration of the Independence of Cyberspace" in: https://www.eff.org/es/cyberspace-independence

be about creativity, playing, collaboration, and participation. The ideas associated with cyber-culture were, and continue to be, aligned with those discourses that argued that the internet and its technologies would create a new type of society that would overcome the problems of the existing world (Bory 2020, Breton 2011). Although more than two decades have passed since these postulates about the development of the web and the benefits it would bring to all human beings, these narratives continue to be valid and have permeated daily and specialized discourses.

However, when we refer to algorithmic culture, we are not thinking of a new culture, much less an improved culture free of tensions, problems, and power relations. Algorithmic culture is one in which algorithms increasingly condition what users see, read, and hear in online environments. This means that today the algorithms of Facebook, Instagram, Twitter and many other applications are the ones that decide what we consume, to which we dedicate time and attention, as a result of "our taste and daily experiences —such as friendship, entertainment, information— are crisscrossed by algorithms that have ended up colonizing our most intimate spaces, modifying tastes and interests" (García and Valle 2020).

Access to culture, consumption, leisure time, and a relationship with the world are increasingly mediated by algorithms, which shape and condition our online experience.

> *"Users do not simply consult websites or talk to their friends online. Social media and other commercial Web companies recommend, suggest, and provide users with what their algorithms have predicted to be the most relevant, hot, or interesting news, books, or movies to watch, buy, and consume" (Bucher 2018).*

As explained in the previous section, the algorithms carry the DNA of their creators, and in the case of the algorithms of the applications (apps) and services we use, they have the DNA not only of their creators but also of the companies to which they belong. Because of this, the algorithms of Facebook, Google, Uber, Netflix, and Tinder are oriented to their interests and business models, not necessarily to the users' needs.

No matter how powerful the discourse is that these applications and services want to make our lives easier, simpler, and more efficient, their algorithms are designed to occupy and capture our time and preferences. The objective of the mentioned companies is that we spend more time on their applications, not that necessarily receive better information or the shortest path or the indicated date. Instead their motive is to collect user data to improve their services, while also exploiting them (Williams 2021, Zuboff 2020).

We might think that algorithms act and intervene in mundane issues: the series and movies that we watch on Netflix or YouTube, the news and images that we see on Facebook or Twitter, the songs that we listen to on Spotify, the route we take to get from one place to another or how much we should pay for a plane ticket. But the truth is that algorithms increasingly participate in fundamental matters and processes of individual and collective life such as justice, work, health, social welfare, and security, among others, so their cultural significance and importance is evident.

> *"Algorithms play a central, if obfuscated, role in society because they rank, sort, discriminate, predict, and rate all aspects of life. [...] Algorithms perform tasks related to encrypting, routing, ranking, filtering, searching for, and compressing information—permeating every aspect of information and communication technology infrastructure. They also make consequential decisions related to how to value, score, or tailor political speech to people online" (DeNardis 2020).*

Currently, it is impossible to deny that algorithms are leaving marks on the daily lives of peoples and societies by shaping and conditioning users' behaviour. There is hardly an activity, space, or moment in which the algorithms of applications and technology companies do not intervene. Due to their cultural importance, algorithms must be studied, understood, and problematized. Blind and unrestricted trust in its procedures and results must be overcome.

Algorithms not only want us to spend more time with them in algorithmically mediated environments, but the desire of their programmers is also to influence everyone's daily lives, since what occupies our time and attention conditions and influences what we do and think (Patino 2020, Williams 2021). Algorithms influence and persuade our behaviours and meanings about reality and the world.

It is not only that we now develop more activities digitally, but digital is conditioning what we do in the physical world.

> *"Math makes possible all these algorithms that have come to invade and almost run our lives. For centuries, math was something we drew on in making observations about our world. Now it is a potent tool we use to shape our planet, our lives, and even our culture" (Steiner 2012).*

Consequently, we speak of algorithmic culture because not only have algorithms conquered a good part of cultural consumption, but also because they profoundly influence our worldview, senses, and meanings from which we relate and interpret our close and far away surroundings.

Final Reflection: Limits of Algorithmic Culture

Algorithmic culture is a culture based on the calculation that, beyond making our lives easier and more efficient, it is interested in collecting user data to exploit this commercially. This, in addition to affecting aspects such as privacy and data security (Zuboff 2020) has implications on how users interact in the online and offline world. "Algorithmization of reality, where human life and society are reduced to quantifiable and manipulable data that, rather than predicting or recommending certain behaviours, shape our decisions or choices in different areas of daily life" (García and Valle 2020).

For all of the reasons discussed above, the importance of algorithms is not only about what they are and how they work, but what they do, how they act and what effects and consequences they have on people, their environments and realities. "Computational algorithms may be presented as merely mathematical, but they are operating as culture machines" (Finn 2018).

As noted above, there is no homogeneous and static definition of culture; consequently, when we refer to algorithmic culture we do not want to propose a monolithic definition either; rather, we think of it as an invitation to explore the implications of algorithms in everyday life.

From this perspective, some authors expose the implications that this algorithmization imposes on development and cultural experiences, giving rise to "a monotechnological culture in which modern technology becomes the main productive force and largely determines the relationship between human beings and non-humans, the human being and the cosmos, nature and culture" (Hui 2020). When meanings and comprehension of the world are left in the hands of algorithmic mediation, we fall into a kind of eternal return, where meanings and senses return to us again and again, and therefore it is not possible to see and know beyond what we have previously thought, seen, and believed; a kind of straitjacket that limit cultural experiences and our being in the world. "Algorithmic induction can lead to a kind of [...] determinism, in which our past clickstreams entirely decide our future. If we don't erase our Web histories, in other words, we may be doomed to repeat them" (Pariser 2011).

This is a question that Pariser (2011) has noted in which the consequences of the power of the algorithmic mediation is not an exclusively personal issue, but also a cultural one given that:

> *"There's less room for the chance encounters that bring insight and learning. Creativity is often sparked by the collision of ideas from different disciplines and cultures... By definition, a world constructed from the familiar is a world in which there's nothing to learn. If personalization is too acute, it could prevent us from coming into contact with the mind-blowing, preconception-shattering experiences and ideas that change how we think about the world and ourselves" (Pariser 2011).*

Thus, we could end up caught in a kind of circle that is continuously feeding back with the data and information registered and projected through the calculation and quantification proposed by the algorithm; or in the best of cases, those limits could only be overcome but only in appearance and only according to its own possibilities recommended and allowed by its design. "Data tell what happened but not why it happened. In the absence of causal knowledge, even the best predictions are only extrapolations from the past" (Zuboff 2020).

This could have repercussions on how users communicate, interact, and experience the world, since diversity, imagination, knowledge, learning and openness to new experiences and points of view is circumscribe to the limits pre-set by the algorithms of the platforms and applications; and, therefore, diverse perspectives and opportunities to understand the world and its meanings would not be opened.

> *"Total interconnection and total communication by digital means does not facilitate encounters with others. Rather, it serves to pass over those who are unfamiliar and other, and instead find those who are the same or likeminded, ensuring that our horizon of experience becomes ever narrower" (Han 2017).*

This is not to say that the cultural experiences of human beings have not previously been limited or conditioned by circumstances or institutions of power, as the philosopher Hanna Arendt well expressed; human beings always develop within the natural and artificial conditions created by them.

> *"In addition to the conditions under which life is given to man on earth, and partly out of them, men constantly create their own, self-made conditions, which, their human origin and their variability notwithstanding, possess the same conditioning power as natural things. Whatever touches or enters into a sustained relationship with human life immediately assumes the character of a condition of human existence. This is why men, no matter what they do, are always conditioned beings. Whatever enters the human world of its own accord or is drawn into it by human effort becomes part of the human condition. The impact of the world's reality upon human existence is felt and received as a conditioning force" (Arendt 2019).*

However, it does seem that these limits are now more stressed, as a kind of cultural determinism marked by algorithms. So, algorithmic culture, which would be the bearer and creator of meanings in contemporary life, would be reduced to the interest of technology companies, the values of their designers and programmers – "white men geeks" and "venture capitalists" – who have achieved a cultural dominance that is imposed on the rest of societies (Couldry and Mejias 2019, Lovink 2019).

The role of ICT users would be reduced to mere consumers since the content they receive and their interpretation of reality are fed back from their own ideologies and beliefs. In this way, any ideas, content and interpretations that they do not share will disappear from their view, and the consumers-users will hardly be aware that they exist. Thus, "neoliberalism makes citizens into consumers. The freedom of the citizen yields to the passivity of the consumer" (Han 2014). This would call into question, among other things, the idea of digital citizenship, since being a citizen implies seeing things from another point of view that is often not the same as ours (Han 2014, Pariser 2011), which is conclusively relegated to the background under the dominance imposed by algorithms.

This drift to which users are subjected, instead of expanding their frameworks and cultural references, ends up restricting them. As Lovink (2019) pointed out "the multiplicity of sources and points of view, once celebrated as "diversity of opinion", is now reaching its nihilistic "zero point", which, like it or not, has consequences on cultural training and experience.

Finally, we return to an idea of one of the great cultural thinkers of the last century, Raymond Williams: the British cultural critic was already warning us that technological determinism inevitably produces a certain social and cultural determinism that "ratifies the society and culture we now have and especially its most powerful internal directions" (Morozov 2011).

All of the above is nothing more than an invitation to reflect on the cultural frameworks that are imposed today in the digital age, in which it is necessary to think about the implications of the algorithmization of cultures and societies, since we find ourselves in "the crucial time for critical theory to reclaim lost territory and bring on exactly this: a shift from quantitative statistics and mapping to the messier, more subjective, but altogether more profound qualitative effects —the incomputable impacts of this ubiquitous formatting of the social" (Lovink 2019), and more specifically in terms of meanings and scopes when we speak of algorithmic culture.

The transdisciplinarity of the social and human sciences provides the intellectual resources to understand algorithms beyond their technical characteristics and begin to understand their social implications. The idea of algorithmic culture is not proposed as a new academic fashion or is employed to trivialize the concept of culture that has been used to refer to everything (culture of violence, organizational culture, youth culture, among others). With this new term, we want to draw attention to the implications of algorithms on people's daily lives; in our way of seeing, understanding and relating to the world. It is not another culture, like cyber-culture, it is a part of the culture.

References

Arendt, H. 2019 [1958]. *La Condición Humana.* Barcelona: Paidós.

Berlinski, D. 2000. *The Advent of the Algorithm. The Idea that Rules the World.* New York: Harcourt.

Bory, P. 2020. *The Internet Myth: From the Internet Imaginary to Network Ideology.* London: University of Westminster Press.

Bucher, T. 2018. *If... Then. Algorithmic Power and Politics.* New York: Oxford University Press.

Breton, P. 2011. *The Culture of the Internet and the Internet as Cult*. Duluth: Litwin Books.
Broussard, M. 2018. *Artificial Unintelligence. How Computers Misunderstand the World.* Cambridge, Massachusetts: MIT Press.
Cardon, D. 2016. Deconstructing the algorithm. Four types of digital information calculations. *In*: J. Roberge, and R. Seyfert, (Eds.), *Algorithmic Cultures. Essays on Meaning, Performance and New Technologies* (pp. 95–110). New York: Rutledge.
Cardon, D. 2018. *Con que sueñan los algoritmos. Nuestra vida en el tiempo de los big data*. Madrid: Ediciones DADO.
Couldry, N. and U. Mejias. 2019. *The Costs of Connection. How Data is Colonizing Human Life and Appropriating it for Capital*. Stanford: Stanford University Press.
Couldry, N. 2020. Recovering critique in an age of datafication. *New Media & Society,* 22(7), 1135–1151. DOI: 10.1177/1461444820912536
Crosby, A. 1998. *La medida de la realidad. Cuantificación y la sociedad occidental*, 1250–1600. Barcelona: Crítica.
DeNardis, L. 2020. *The Internet in Everything. Freedom and Security in a World with no Off Switch*. 79, 133–134. United States of America: Yale University Press.
Domingos, P. 2015. *The Master Algorithm: How the Quest for the Ultimate Learning Machine will Remake our World*. New York: Basic Books.
Eubanks, V. 2021. *La automatización de la desigualdad. Herramientas de tecnología avanzada para supervisar y castigar a los pobres*. Madrid: Capitán Swing.
Finn, E. 2018. *La búsqueda del algoritmo. Imaginación en la era de la informática*. Barcelona: Alpha Decay.
García, D. and D. Valle. 2020. Los impactos de la ideología técnica y la cultura algorítmica en la sociedad: una aproximación crítica. *Revista de Estudios Sociales*. 71, 15–27. https://doi.org/10.7440/res71.2020.02
Geertz, C. 2003. *La interpretación de las culturas*. Barcelona: Gedisa.
Gillespie, T. 2016. #trendingistrending. When algorithms become culture. *In*: J. Roberge, and R. Seyfert (Eds.), *Algorithmic Cultures. Essays on meaning, performance and new technologies* (pp. 52–75). New York: Routledge.
Han, B.C.H. 2014. *Psicopolítica*. Barcelona: Herder.
Han, B.C.H. 2017. *La expulsión de lo distinto*. Barcelona: Herder.
Harris, M. 2001. *Antropología Cultural*. Madrid: Alianza Editorial.
Heidegger, M. 2003. *The End of Philosophy*. The University of Chicago Press.
Hui, Y. 2020. *Fragmentar el futuro: ensayos sobre tecnodiversidad*. Buenos Aires: Caja Negra.
Johnson, S. 2012. *Futuro perfecto. Sobre el progreso en la era de las redes*. Madrid: Turner.
Kottack, P. 2011. Antropología Cultural. México: McGraw-Hill
Kuper, A. 2001. *Cultura: la versión de los antropólogos*. Barcelona: Paidós.
Kroeber, A. and C. Kluckhohn. 1963. *Culture: A critical review of concepts and definitions*. New York: Vintage.
Lévy, P. 2007. *Cibercultura. La cultura de la sociedad digital*. España: Anthropos.
Lohr, S. 2015. *Data-ism*. New York: HarperCollins.
Lovink, G. 2019. *Sad by Design. On Platform Nihilism*. London: Pluto Press.
Manovich, L. 2017. Los algoritmos de nuestras vidas. *Cuadernos de Información y Comunicación*, 22, 19–35. https://doi.org/10.5209/CIYC.55960
Mayer-Schonberger, V. and K. Cukier 2013. *Big data. La revolución de los datos masivos*. Madrid: Turner.
Morozov, E. 2011. *The Net Delusion*. Public Affairs.

Morozov, E. 2015. *La locura del solucionismo tecnológico.* Madrid: Clave intelectual.
Noble, S. 2018. *Algorithms of Oppression. How Search Engines Reinforce Racism.* New York: New York University Press.
O'Neil, C. 2018. *Armas de destrucción matemática. Cómo el big data aumenta la desigualdad y amenaza democracia.* Madrid: Capitán Swing.
Patino, B. 2020. *La civilización de la memoria de pez. Pequeño tratado sobre el mercado de la atención.* Madrid: Alianza editorial.
Pariser, E. 2011. *The Filter Bubble.* New York: The Penguin Press.
Rid, T. 2016. *Rise of the Machines: A Cybernetic History.* New York: Norton & Company.
Rosa, H. 2020. *Uncontrollability of the World.* Cambridge: Polity Press.
Sadowski J. 2019. When data is capital: Datafication, accumulation, and extraction. *Big Data & Society*, 1–12. DOI: 10.1177/2053951718820549
Sfez, L. 2005. *Técnica e ideología. Un juego de poder.* México: Siglo XXI
Srnicek, N. 2018. *Capitalismo de plataformas.* Buenos Aires: Caja Negra Editora.
Steiner, C. 2012. *Automate This: How Algorithms Took over Our Markets. Our Jobs, and the World.* New York: Portfolio/Penguin.
Stephens-Davidowitz, S. 2019. *Todo el mundo miente. Lo que Internet y el big data pueden decirnos sobre nosotros mismos.* Madrid: Capitán Swing.
van Dijck, J. 2014. Datafication, dataism and dataveillance: Big Data between scientific paradigm and ideology. *Surveillance & Society*, 12(2), 197–208.
van Dijck, J., Poell, T. and de Waal, M. 2018. *The Platform Society.* New York: Oxford University Press.
Williams, J. 2021. *Clics contra la humanidad. Libertad y resistencia en la era de la distracción tecnológica.* Barcelona: Gatopardo Ediciones.
Winner, L. 1980. Do artifacts have politics? *Daedalus*, 109(1), 121–136.
Zuboff, S. 2020. *La era del capitalismo de la vigilancia.* Barcelona: Paidós

CHAPTER

2

The Functioning of Attention as a Behavioural Prediction Mechanism in Social Media

Pedro Nicolás Aldana Afanador
Department of Social Communcation, Autonoums University of Bucaramanga
Av. 42 # 48 - 11, Bucaramanga, Santander, Colombia

Introduction

Why do we use 'pay' attention rather than other verbs such as: give, address, provide or to be attentive? One of its first references dates from the year 1737 in 'The Political State of Great Britain', volume 53, and after many other appearances in newspapers of the time, the Samuel Johnson's dictionary (1755) included the modern phrase 'pay attention' to express a meaning for regards. According to the Oxford English Dictionary, to pay means "to render, bestow, or give". This etymological heritage from the Romans, from the verb *pax*, which eventually became pay, only found a privileged place in the 18th century.

However, one of historical curiosities about the appearance of this modern phrase is that its boom occurred simultaneously with the birth of the modern circus in London and United States. It seems that the nascent entertainment industry had an intimate relationship with its acceptance and its use in everyday language. The great peculiarity is that the entertainment industry lives off the 'paying attention' of each one of us. Where the reflector points is where our gaze is set. The show is designed to capture this human ability, our attention.

The entertainment industry, however, has not been the only one that has taken advantage of human attention. In a quickstudy, advertising, and marketing have done it as well. Currently, a large new sector works with the 'pay attention'. Since the 1960s, we can find a new concept that analyses and describes the relation

Email: paldana58@unab.edu.co

between attention and economics: the economy of attention. The first author to propose this correlation was Simon (1969) who observed that:

> *"(...) in an information-rich world, the wealth of information means a dearth of something else: a scarcity of whatever it is that information consumes. What information consumes is rather obvious: it consumes the attention of its recipients. Hence a wealth of information creates a poverty of attention and a need to allocate that attention efficiently among the overabundance of information sources that might consume it".*

Consequently, the economy of attention changes the initial question: Why do we say 'pay attention?', and suggests a new inquiry: Is it possible to pay for human attention? Let's consider an example: In the 2018–2019 edition of the Champions League, after a thrilling quarter-finals and semi-finals, two English teams went to the Wanda Metropolitano in Madrid to fight for the cup. As the hard-fought match was being played, model Kinsey Wolanski jumped onto the pitch distracting the fans watching the game. In a premeditated act, she ran towards the centre of the field chased by two guards to promote her partner's brand.

As a result, this 45–second pitch invasion stopped the match and caught the attention of spectators who, in turn, made the act viral. Due to television production protocols, the act was not broadcast live; instead, the photos and videos taken by the spectators' mobile phones quickly went around the world. The model, who had three hundred thousand followers on Instagram, gained two and a half million followers in less than 24 hours. An average of one thousand five hundred followers per minute. In addition, her appearance on the field was valued at around three million pounds in advertising and marketing.

Given this case, is it possible to pay for human attention? Yes, it is possible. Since the consolidation of web 2.0 and digitalisation processes, social media platforms have positioned themselves exponentially in everyday life. We often believe that they are free of charge. Social media makes money at the expense of our attention span. According to the Digital 2021 report (We Are Social 2021), we spend 6 hours and 54 minutes a day consuming the internet, where 2 hours and 25 minutes are spent on social media (We Are Social 2021). This means that we are investing more than two days of work for these industries in a week, which is approximately £178 for our time invested.

Social media is the main niche of the attention economy. These platforms are designed to capture and hold our attention for as long as possible. This consumption time is proportional to the attention capital invested. In other words, the more time we spend consulting our social media, the greater the attention capital invested in it, which translates into profits for the digital company. Thus, the concept of attention changed to digital attention. The big difference lies in the profit margin that the new technological-communication industries can generate, placing human attention at the centre of the business model's value chain.

For all these reasons, this attention today becomes a treasure that we must safeguard and understand. Moreover, the attention economy becomes a key to understand this reality, and to analyse social media from a more critical perspective. In this second chapter we lay the groundwork for understanding how human attention works and how it relates to social media networks. As predictors and inducers of digital attention consumption, social media remains at the core of our behaviour. Let's move beyond why we say we 'pay' attention, to understand why social media make money thanks to our attention.

What Is the Human Attention?

Attention is one of the most valuable resources we possess, but few people are aware of it. We must understand that it is not easy to talk about attention. We have always heard of it, but it has been rarely defined for us. To make an analogy, attention is the gateway to our brain process. It is like a host that addresses all the inputs we receive. Therefore, three elements are immersed in the attention process. They are essential to reach a definition: Orientation Reflex, Selection and Concentration.

If you have ever been in a cafe and suddenly turned around because you heard a familiar voice. Or your mobile phone rings and you automatically look at it. Or perhaps you meet the person you like unexpectedly in the street, and you start sweating and your heart starts beating faster. In these cases, you have experienced the orientation reflex.

This reflex is an immediate response produced by a perceived stimulus. As a result of this input, our brain responds with what is known as a peripheral response. This reaction, on the one hand, can be somatic, such as: stopping or inhibiting the action or task we are performing; or changes in the orientation of our head and body posture. And on the other hand, the vegetative type, which is more linked to the galvanic response as an electromagnetic reaction of our body; changes in the rhythm of the heartbeat; and dilation of the blood vessels, among others.

According to Sokolov (1963), the attentional response to a stimulus is directly proportional to the magnitude of the stimulus received. In other words, if our attention were like a capital that we invested in the different stimuli we receive, the one that has the greatest impact would have the greatest attentional capital. Two of the main characteristics of stimuli are intensity and novelty. The former depends on the composition of the stimulus itself, our expectations, among other variables. For example, elements that have an intimate relationship to us, like a marriage proposal, an anniversary; or unexpected news such as a generous job offer, the death of a family member, an accident, etc.

The latter implies that a repetitive stimulus generates what will be called habituation. The novel stimulus will be the one that has the greatest impact on our attention. It is not the same to greet and get to know a person we see every day as it is to greet a special person whom we do not usually meet. A first date or a first kiss has a greater impact. Stimuli will determine our orienting response and the intensity of the attention we pay to them. The more noticeable a stimulus is, the more our attention will be directed towards it.

The second key element in understanding attention is selection. How do we choose a stimulus? Several theorists have attempted to define a model of information processing. For example: Broadbent (1958), Deutsch and Deutsch (1963), Treisman (1964), Atkinson and Shiffrin (1968), Newell and Simon, (1972), Atkinson and Juola (1974), Massaro and Oden (1978), and Johnston and Heinz (1978), among others. In a simplified form, the models essentially describe a stimulus or input that is perceived by our senses or sensory memory. Then, it passes through a filter or analyser, where a decision is made on whether to make it conscious or not. This filter has a limited processing capacity and can undergo attenuation or noise actions. Finally, our brain responds or generates an action to the perceived stimulus.

Let us take an example: You are driving along, talking on the loudspeaker with your office colleague, when you come around a corner and see a ball cross in front of you. Our sensory memory, accompanied by the orienting reflex, automatically focuses our attention on finding a quick response to the event, and cancels all the activities we were doing. Therefore, the automatic reaction in the attentional selection exercise is to slow down the vehicle, as we assume that a person may be coming behind the ball. Another example is when several people speak to us at the same time, or we receive several stimuli, our attention has the capacity to select one.

Selective attention is the basis for the coordination and control of human actions. Attention helps us to discriminate and prioritise stimuli in order to respond to them. Hence, this presents us with two peculiarities: firstly, the ability to process stimuli and select them is a mental and physical task at the same time, which will lead to fatigue or exhaustion in the long run. Secondly, not all perceived stimuli are attended to. There are many more stimuli that we reject in our daily lives. Selective attention is closely related to human behaviour. Attention helps us to organise our behaviour, which will make it the first step in any action we take.

The third element is sustained attention or concentration. We can define it as the ability to maintain attention on a specific action for a long time. This ability is the result of the selection of a stimulus, which we maintain over time. Some studies show that the attention cycle oscillates in periods of fifteen minutes. Other studies tell us about the theoretical curve of fluctuations of attention that will always have a higher level at the beginning and at the end of the process. In general, all elements of attention can be trained; therefore, it can be modified for or against the attentional cycle.

For example, a student will have a longer attention span than a person working on a production packaging line. Just as a bus driver will have a longer attention span than a cashier in a shop. Clearly, the interaction of stimuli will have a bearing on the level of concentration with which the task will be performed. In addition, another relevant factor in selection is the level of motivation with which we respond. So, the best synonym for attention is motivation.

Attention is a multidisciplinary term that brings together various relationships and processes. A first definition is that it is a scarce resource. Our attention is limited, hence it is of great value, as mentioned above. However, the Oxford dictionary defines attention as "the act of listening to, looking at or thinking about something/someone with attention; the interest people show in someone/something". It presents the concept as an action. However, two branches of knowledge such as neurophysiology and neuropsychology can serve to better clarify this complex concept.

From neurophysiology, attention can be defined as the amplification or increase of neural activity in a particular brain area involved in processing a stimulus, which allows selective focus on a task and is essential for a series of actions linking several cognitive functions. Attention can be considered as "the interface between the large number of stimuli coming from our complex environment and the limited set of information of which we are aware" (Rueda et al. 2015).

As Raz (2004) proposes, "researchers in the field agree that attention is not a unitary term. Rather, we can break attention down into more circumscribed subsystems of function and anatomy". Moreover, "attention is a complex neurobehavioural capacity without which the expression of all other higher functions of the human brain is impossible" (Filley 2002). Thus, attention will be approached from the integrative theory formulated in the 1990s by Posner and colleagues (Posner and Petersen 1990, Posner and Rothbart 1991, Posner and Dehaene 1994) called the Three Networks. "This theory argues that the variety of attentional manifestations is produced by separate but interrelated attentional systems" (Fuentes and Lupiáñez 2003). The model is composed of Alertness, Orientation and Executive Function.

In addition, from a neurophysiological perspective, there are two attentional processing mechanisms that enable this capacity to be activated. The first is called top-down, which "represents the selection processes directed towards particular goals, which produces a greater neuronal activation of the relevant sensory input to discriminate the stimulus of interest from those that are not relevant to achieve the goal" (Ruiz and Cansino 2005). And the second is called bottom-up, which "is associated with the processes involved when attention is directed to a specific stimulus because certain characteristics of the stimulus stand out, such as its infrequency, novelty, intensity or contextual relevance" (Ruiz and Cansino 2005).

These two mechanisms can be illustrated as follows: Firstly, when we are watching a Netflix series that captures our greatest attention capital and we are so focused that we do not hear any calls or information given to us. Or when we are so busy with a task that time passes quickly without us noticing. Secondly, it is an instantaneous reaction, for example, when an activity is in progress and a notification appears on our smartphone, so we change our focus of attention. Or we see that the traffic light turns red, and we automatically slow down and apply the brakes of our vehicle.

In terms of neuropsychological studies, as explained by García Sevilla (1997), attention can be understood as "a mechanism that sets in motion a series of processes or operations thanks to which (...) we are more receptive to the events of the environment and allows us to perform numerous tasks more efficiently". From this perspective, attention is subject-dependent and therefore unique and varied. "There are individual differences in our attention span. However, one of the most important characteristics of these strategies is that they are not innate but learned. This is important to consider, not only because they can be modified and improved with practice, but also because it is possible to develop strategies aimed at improving the functioning of the different attention mechanisms" (García Sevilla 1997).

It should be noted that attention for neurophysiology and neuropsychology is a process in which different elements are involved at the same time; these are our nervous, endocrinological, muscular and vascular systems in a single instant. It is a resource that focuses the entire body and mind (or psyche) on the stimulus that has captured our attention. It is important to remember that attention is a complex system in which there are many variables and adjacent functions. Although neuroscience has studied its composition and mechanisms, it is an evolving concept. To talk about attention is to talk about something complex, because it is a multidisciplinary term that brings together several concepts. But to define it simply: attention is a scarce resource, so the attention economy is the management of the scarce resource of human attention.

What Are the Main Characteristics of Our Attention?

Having defined the concept, there are five main characteristics that are important to know to further understand human attention. These are uniqueness, breadth, change, intensity, and control. The first characteristic is a reminder that attention is unique. Each person has a preferred way of paying attention. We are often wrong to demand that others pay attention to a stimulus in the same way as we do. For example, we have met distracted people who find it difficult to concentrate when listening to a speech, attending a class, writing a paper, but are able to attend to and learn several movements for a dance, or memorise a script for a play perfectly. Or people who we think are unfocused, but when they build a creative process or develop an activity, they can do it at an impressive speed. Our sensory memory works in a distinctive way and picks up stimuli from the environment in a particular way. It is important to note that if the stimulus proposes a greater involvement of the senses, our response will be greater.

In this context, this characteristic leads us to ask ourselves about inattention, which also marks the uniqueness of each person. The opposite of attention is distraction or dispersion, for example: the difficulty of maintaining attention on a task, action, or information. For this type of person, any new stimulus generates a change in attention, which will make it difficult for them to concentrate. In addition, there are various diseases that affect the attentional process, for example, sleep disorders, epilepsy, attention deficit disorders, genetic disorders that directly affect this process such as Down, Williams or Tourette and personality disorders, among others.

The second characteristic is breadth. Certain activities require paying attention to more than one action at the same time. This is commonly known as multitasking. When you look at the amount of information or the number of possible responses at the same time, it is breadth. This increased responsiveness occurs because we can combine different types of attention. Several authors have tried to determine and classify types of attention. Today, there are different typologies, but commonly we find selective, divided, and sustained types of attention.

On the other hand, according to Sohlberg and Mateer's hierarchical model (1987, 1989), attention can be of the following types: a) Internal when it is focused on mental processes, b) External when the inputs come from the environment, c) Overt when it is accompanied by motor responses, d) Covert when we attend without a specific physical response, e) Aurosal which refers to the level of activation and brain response, f) Focalised when it focuses attention on one stimulus, g) Sustained when it maintains attention on one stimulus for a long time. h) Selective when it chooses and receives one stimulus among many others, i) Alternating when it changes the focus of attention on two or more inputs. j) Divided when it attends to different stimuli at the same time, k) Auditory when it perceives stimuli through the ears, and l) Visual when it perceives stimuli through the eyes.

Within these characteristics, it is possible to discover that attention is modifiable due to the brain property called neuroplasticity. For example, when we are learning to drive and someone speaks to us, our attention is divided, and we find it difficult to respond to other stimuli. Gradually, practising this activity helps us to incorporate new tasks with attention. Thus, neuroplasticity is the brain's capacity to adapt to changes by creating new connections that allow us to modify our attentional response. There are exercises to improve the brain's response to different stimuli or, on the contrary, to reduce this property.

The third characteristic indicates that attention is changeable. Linked to the previous characteristic, it proposes the phenomenon of shifting, which means a jumping of attention to different tasks to respond to different stimuli. This occurs because there are different types of attention to combine at the same time. For example, we can drive the car, drink water, and sing while listening to the radio. We can also answer a phone call while still working on the laptop and following our favourite team's game. Or we can dance and talk at the same time, among many other combined actions. Attention can respond to more than one stimulus at the same time, known as divided attention.

Consider, for example, the famous cocktail effect. When we arrive at a party or a restaurant, we tend to focus our attention on the conversation with our friends. While this is happening, we are receiving different auditory stimuli from other diners or tables in the place. However, we can filter out the other inputs to concentrate on the conversation despite the noise in the room, and simultaneously, we are able to follow the conversation of another table or person at the cocktail party.

This feature recalls Kahneman's (1973) model of the attentional capital that can be invested in different tasks, and the responses to each of them. This theorist proposes that attention is a capital that is invested according to the magnitude of the stimulus received, as discussed in this chapter. By dividing the capital between different stimuli perceived at the same time, the attentional response will not have the same efficiency. If we divide attention to respond to many activities at the same time, our efficiency in responding to these tasks will suffer. Consequently, the time spent responding to the processed stimulus increases. For example, checking social media while writing a chapter of a book generates more time spent on the writing process. If attention is shifted to multitasking, more time will be spent.

As a result, this capacity is shown to be limited. This principle is the foundation of the economy of attention. If we are reading this book, we cannot be playing a basketball game. Or if we are listening to a lecture, we cannot be playing video games at the same time without our attention being affected. As much as we want to pay attention to all stimuli at the same time, it is essential to focus on just one, because the effectiveness of your response will not be the same.

The fourth characteristic is intensity, which is related to responsiveness. The direct relationship between stimulus magnitude and attentional response was presented earlier in this chapter. Intensity indicates the amount of information that can be attended to at the same time and the number of tasks that can be performed simultaneously. In other words, the level of attention with which we respond. It is important to note that attention is fluctuating, so responses will not always have the same intensity. Sometimes, we have felt that we got off on the wrong foot. If we talk to our friends, we do not understand their jokes; or we open the fridge and do not remember what we were looking for; in general, we feel distracted. Attention does not always respond in the same way. This situation can also be related to physical or mental tiredness, drowsiness, emotional or personal situations, etc.

The fifth characteristic is control. This is one of the most important properties of attention. Imagine for a moment that you respond to all stimuli at the same time without any filter. Attention plays a fundamental role in discriminating and organising the tasks we are going to perform. The control path of automation is one stimulus, one route, one response. For example, when we are in danger, our brain responds in a specific way. Or when we must make a quick decision, our brain decides.

Attention controls how the human brain approaches and selects information; it guides exploration and search processes; it controls the ability to concentrate over time; it suppresses distractions that affect attentional processes; it discards inappropriate responses; and despite being tired or bored, it helps us to maintain our attention (García Sevilla 1997). It is important to remember that attention has a unique way of functioning. There will always be a first moment of initiation, which is when the stimulus is consciously or spontaneously perceived and selected. Then, a second moment of sustaining, in which attention is maintained to respond. Finally, there is a third moment of termination when attention wanes and the process restarts.

As for the last two attentional moments, there can be spontaneous recovery of attention, which are the moments in which the system refocuses on a stimulus despite having lost sustained attention (García Sevilla 1997). For example, when we are using Instagram for a long time and we keep swiping through photos or stories but we are not paying attention until we suddenly become aware of the action we are performing. Or when we are in class and for a moment, we get bored or lose attention, but after a while we regain focus.

In conclusion, attention plays a fundamental role as a gateway and selection of stimuli found in the environment, which means that neuro-physiologically and neuro-psychologically it is of great value for states of consciousness and for our behaviour. There are also different types that can be combined to respond to several perceived stimuli at the same time. But the more attention is divided, the less efficient it becomes, making it a scarce resource for human beings. Consequently, attention is a function that has a limiting characteristic, since the human brain can only concentrate on a certain number of actions at a time, a characteristic that the economy of attention will make the most of.

Within its complexity, attention plays a strategic role in our lives and social media knows this and uses the attention economy to its advantage. This will lead to a shift in our approach to attention. Seen as a commodity, attention takes all its characteristics and fundamentals to control our behaviour and seeks to extend our consumption time on these platforms, which will result in the profound impact of social media on our attention. The attention economy opens the door to the era of digital attention.

How does Social Media Capture Our Attention?

Social media are designed to control our attention. The economic conceptualisation of this scarce resource and its implications for our behaviour establishes a relationship between the functions of attention and the exercise of digital consumption. By understanding what our attention is and how it functions, it is possible to make a critical reading of the mechanisms that social media uses to make money from our attention. In the attention economy the new currency is our time, and these platforms generate an environment of personalised attention to extend our consumption time every day.

Moving to digital attention, functionally and structurally, social media has a sophisticated way of controlling our attention. The new attention industries have developed four mechanisms that have a direct effect on our brain, as shown in the table below:

Table 2.1: Mechanisms of Social Media Operation and Effects on Human Attention

Mechanism	Stimulus	Affectation	Effect
Notifications	Systematic and overabundant auditory, visual, and sensory impulses	Attentional network saturation	Over-alert state Anxiety
Messages and Posts	Audio-visual frequent, short, simple, diverse, interactive, and dynamic data	Constant attention breaks	Attentional dispersion
Fear of Missing Out (FoMO)	Apprehensive graphic and functional environment, and flow of information and interaction from the selection of topics and people of interest to the user to create a filiation relationship	Dependency	Anxiety Stress
Likes and Rewards	Social acceptance stimulus that reinforces the behavioural character resulting from the user's interaction or publication	Assimilation and normalization of repeated consumption behaviours	Addiction

Source: Giraldo-Luque, Aldana and Fernández-Rovira (2020)

Notification: Over-alert to the Brain

Think for a moment about the number of notifications we receive every day, and which reach us via email, or from our social media networks on the screen of our mobile phones or smart-watches. Today we receive an overabundant number of alerts or stimuli that saturate our attention networks. Therefore, notifications have become systematic impulses that are designed to activate our bottom-up mechanism. We look daily at our smartphone screens, hoping to find a notification.

In their setup, notifications are designed to capture our attention through vibration, visual stimulus such as pop-ups, a personalised sound, special codes and/or bright colours. These digital stimuli constantly activate our orientation reflex or alertness network (Posner et al. 1990, 1991, 1994) providing a renewed sense of vigilance.

The social media notification system uses surprise, novelty, and repetition to generate neurophysiological changes that affect our behaviour. How many times have we changed our activities to respond to a notification, or forgotten our work to check some new information? Notifications act as systematic distractors of our attention. They are push-ups that continually invite us to consume social media, creating self-control problems.

Notifications are a repetitive stimulus with novel information. In other words, the same settings are used, but they always show us novelty and diverse information. If we examine the notifications we receive, most of them handle information that is of little relevance. "Repetition does not represent, in this case, a problem for attention since it generates new and surprising emotions each time" (Giraldo-Luque et al. 2020).

Another great characteristic of notifications is their personalised form, which has a direct impact on our emotional system. It determines the level of intensity of the response to the stimulus received. Who leaves unread a new message from mum or dad, or the greeting from an old friend who has written to us again, or ignores the notification about the new post from the person they love? This affiliation effect determines our prioritisation in response to these stimuli. "The emotional intensity of social network notifications works because it is linked to psychological characteristics such as social valuation, self-image, acceptance, social comparison and recognition" (Giraldo-Luque et al. 2020).

As a result, notifications generate two effects on our brain: Firstly, a state of over-alertness in our attentional network due to the overabundance of stimuli that can cause physical and mental exhaustion. Secondly, anxiety when consuming social media. Notifications produce the latter effect because we certainly enter a certain tension and alarm when discovering their content, and at the same time, we feel the pressure of receiving them. We crave notifications without knowing it. Social media affects and controls our attention with this mechanism, which keeps us 'always on'.

A New Break, a New Message

Another mechanism that social media uses to control our attention are messages. Their design is intimately linked to our sensory memory. Let us remember that, as users, we are prosumers on these platforms. So, we are consumers and producers of information at the same time. A tweet, a Facebook post, an Instagram story or a Tik-Tok are prepared to produce information following the attentional parameters for its creation. Three characteristics structurally define social media messages: firstly, they are short and dynamic. Secondly, they are changeable and interactive. Thirdly, they are produced using audio-visual language.

The form and composition of social media messages at the macro level is designed to spend the minimum amount of time consumed in our media diet to amplify the maximum possible amount of post consumption. For example, Instagram or Tik-Tok posts range from fifteen seconds to one minute. However, Twitter might best illustrate this first characteristic for us. A tweet is a dynamic form of microblogging that allows us to share ideas. It started out as a 140-character message that has now doubled in length, and its short form and active use allow us to focus our attention in a more direct way. Twitter is easy and intuitive to use to create a post. In addition, its low learning curve for generating messages in the app gives us the greatest capacity to build content for the platform.

The second feature proposes that messages are changeable and interactive. In the early days of Facebook, we only wrote text messages on the platform, but today it is possible to publish a personalised message by editing the background, the font, adding image, sound, or video, among many other applications that help us build the perfect Facebook post. Social media generates constant changes in the form and composition of its messages, to maintain the novelty for the consumption and production of information.

These constant updates to the platform's interface are designed to avoid the attentional effect of habituation that repetitive use produces, so that we do not get bored. In addition, the publications have incorporated new features to create greater interactivity with each content, for example, questions, polls and voting boxes; integrated image filters, gifs, hashtags, location, web links and a series of elements that seek engagement with the platforms.

The last element is the audio-visual format. Social media messages seek to capture our greatest attention capital and, at the same time, to be a high-intensity stimulus for our brains. Thanks to the audio-visual format, social media captures two important types of attention: visual and auditory. The more intense the stimulus, the more sensory memory we need to respond to it. The audio-visual format will be the key tool for social media, such as Tik-Tok and Instagram, as platforms that have quickly positioned themselves in the attention market.

The structure and functioning of messages have a direct impact on our attentional process, especially in selection. Being short, dynamic, changing, and interactive, they generate constant jumps in our attention. Due to the overabundance of stimuli, the selection process will be increasingly forced to change. A new message is a new break in our attention. If our attention is modified to constantly shift from one message to another on social media, we will adapt our brain to select in the same way, leading to attentional dispersion.

The Fear of Missing Out (FoMO)

If we have ever checked our social media to see what our friends and family are up to; read the latest news; or feel a part of daily events around us; we have experienced the Fear of Missing Out (FoMO). "The attention-grabbing power of social media has generated in users the feeling that, if they do not constantly check their platforms, they will miss something important in their lives" (Giraldo-Luque et al. 2020).

Hooking these platforms to our attention causes a double action to take place in our role as prosumers: we are afraid of not being included; and we are afraid of missing out on something. The first fear will require us to continuously create content to feed our social media networks and generate links with our followers. The second fear is related to consumption. If we do not know the latest trending video, the viral meme, the latest trending topic, or if we do not know about a friend's latest trip, the new status or photo of our colleagues at the office, it seems that we have no social life.

This instilled fear has an impact on our attention control that affects our behaviour, due to the dependency created by social media to respond to their stimuli (notifications, messages, and posts). This may create an addiction to their consumption, which in our attentional process leads us to prioritise the stimuli of these platforms. Consequently, FoMO makes us experience anxiety and stress, since, if we do not post or consume, we stay away from social media, which causes nervousness, distraction, worry, and pressure, which in turn generates an impact on our behaviour.

One Like, One Reward

The last mechanism that social media has to control our attention is the reward system. How many of us have posted on our feed looking for a like or a comment, or have invested more time on these platforms looking to increase our followers, or even paid for advertising to get more attention to your profile? The social media reward system is the perfect secret weapon to capture our attention.

As in the behavioural theory of Pavlov (1928), Watson (1947) and Skinner (1975), the rewards received reinforce our behaviour. A like or a comment is a reward that acts as an action-reaction that dominates our behaviour and supports social media mechanisms. The interactive sense of the social media reward system includes a dual pathway of giving and receiving, as active users have the ability to provide a reaction that will reinforce another user's attentional response, but equally, to receive a positive stimulus that approves and normalises our behaviour.

This system has a direct impact on our emotions. The more followers, comments and likes, the more intense our attentional and emotional response. Therefore, social media platforms are the marketplace for attention. Voluntarily or involuntarily, we seek to capture the attention of other users, which will generate a status based on gratification and self-image.

Another element that supports this system is that the rewards are completely quantifiable. Social media is driven by numbers and statistics and, without realising it, we are part of them. The ability to numerically formulate measures of interactions and their fluctuation makes the reward system a predictive element that helps to control the digital attention of users. In fact, it is possible to find people who are experts in increasing rewards on profiles by generating strategic plans to do so, which means that our attention can be manipulated.

On the other hand, likes also have another special implication in the system, not only reinforcing and normalising our behaviour, but also giving platforms key information about our preferences and choices. The latter mechanism has all our data at its disposal to generate a favourable environment for capturing our attention. A like becomes a piece of information for the algorithm to configure a personalised consumption ecosystem for us and is a valuable input for social media platforms to discriminate the stimuli that attracts our interest.

With the behavioural reinforcements, the direct impact on emotions, the quantifiable rewards and the personalised environment, this mechanism creates a system that feeds directly on us and causes an addiction. We want more likes, followers, and interactions. Excessive use can lead to addiction to a behaviour (Sanchez-Carbonell 2018), such as social media consumption.

Excessive usage can be determined from our loss of control to keep our time on a given action; as a result, we will be more active in substituting basic tasks for social media consumption. This generates withdrawal symptoms, such as anxiety and discomfort about not consuming; produces attentional network tolerance, for example: we will continue to use the platforms even if we are not interested; leads to failure if we try not to use or control our consumption time; and can lead to interpersonal conflicts, as we spend less time with our family and friends.

Social media as predictors of digital attention consumption maintain their mechanisms: notifications, messages, FoMO, and rewards, to control our behaviour. The attention economy shows us that attention is a scarce resource; this is the key to understanding its system. It profits from one of our most important human capacities. The more we understand our attention, the more we can create awareness of the effects of social networks on our brain. So, be aware of the importance of paying attention when you use it.

References

Alter, A.L. 2017. *Irresistible: The Rise of Addictive Technology and the Business of Keeping Us Hooked.* Penguin Press.

Atkinson, R.C. and J.F. Juola. 1974. *Search and Decision Processes in Recognition Memory.* WH Freeman.

Atkinson, R.C. and R.M. Shiffrin. 1968. *Human memory: A proposed system and its control processes. In*: Psychology of Learning and Motivation (Vol. 2, pp. 89–195). Academic Press.

Ballesteros, S. 2014. La atención selectiva modula el procesamiento de la información y la memoria implícita. *Acción psicológica*, 11(1), 7–20.

Beane, M. and R.T. Marrocco. 2004. Norepinephrine and acetylcholine mediation of the components of reflexive attention: Implications for attention deficit disorders. *Progress in neurobiology*, 74(3), 167–181.

Barkley, R.A. and K.R. Murphy. 2006. *Attention-deficit Hyperactivity Disorder: A Clinical Workbook.* Guilford Press.

Broadbent, D.E. 1958. *Perception and Communication.* Elmsford.

Carter, C.S. and M.K. Krug. 2012. Dynamic cognitive control and frontal-cingulate interactions. *Cognitive Neuroscience of Attention*, 2, 89–98.

Corbetta, M., E. Akbudak, T.E. Conturo, A.Z. Snyder, J.M. Ollinger and H.A. Drury. 1998. A common network of functional areas for attention and eye movements. *Neuron*, 21(7), 61–73.

Davenport, T. and J. Beck. 2002. *La Economía de la atención: El nuevo valor de los negocios.* Paidós.

Desimone, R. and J. Duncan. 1995. Neural mechanisms of selective visual attention. *Annual Review of Neuroscience*, 18(1), 193–222.

Deutsch, J.A. and D. Deutsch. 1963. Attention: Some theoretical considerations. *Psychological Review*, 70(1), 80.

Fan, J., B.D. McCandliss, J. Fossella, J.I. Flombaum and M.I. Posner. 2005. The activation of attentional networks. *Neuroimage*, 26(2), 471–479.

Filley, C.M. 2002. The neuroanatomy of attention. *In*: Seminars in Speech and Language (Vol. 23(2), pp. 89–98). Theime Medical Publishers Inc.

Franck, G. 2019. The economy of attention. *Journal of Sociology*, 55(1), 8–19.

Fuchs, C. 2018. *Digital Demagogue: Authoritarian Capitalism in the Age of Trump and Twitter*. Pluto Press.

Fuentes, M.J. and J. Lupiáñez. 2003. La teoría atencional de Posner: Una tarea para medir las funciones atencionales de orientación, alerta y control cognitivo y la interacción entre ellas. *Psicothema* 15, 260–266.

García Sevilla, J. 1997. *Psicología de la atención.* Madrid: Editorial Síntesis.

Giffard, A. 2013. Rhétorique de l'attention et de la lecture. Entretiens du Nouveau Monde Industriel, 17-18 Décembre 2012, Centre Pompidou, Paris, 14-44.

Gerlitz, C. and A. Helmond. 2013. The Like economy: Social buttons and the data-intensive web. *New Media & Society*, 15(8), 1348–1365.

Giraldo-Luque, S., P.N. Aldana Afanador and C. Fernández-Rovira. 2020. The struggle for human attention: Between the abuse of social media and digital wellbeing. *Healthcare*, 8(4), 497. doi:10.3390/healthcare8040497

Giraldo-Luque, S. and C. Fernández-Rovira. 2020. The economy of attention as the axis of the economic and social oligopoly of the 21st century. *In*: S.H. Park, M.A. González Pérez and D. Floriani (Eds.). *The Palgrave Handbook of Corporate Sustainability in the Digital Era*. Palgrave Macmillan.

GWI (2021). Social. https://www.gwi.com/reports/trends-2021

Johnson, M.H., M.I. Posner and M.K. Rothbart. 1991. Components of visual orienting in early infancy: Contingency learning, anticipatory looking, and disengaging. *Journal of Cognitive Neuroscience*, 3(4), 335–344.

Johnston, W.A. and S.P. Heinz.1978. Flexibility and capacity demands of attention. *Journal of Experimental Psychology: General*, 107(4), 420.

Kahneman, D. 1973. *Attention and Effort* (Vol. 1063, pp. 218–226). Englewood Cliffs, NJ: Prentice-Hall.

Massaro, D.W. and G.C. Oden. 1978. Integration of featural information in speech perception. *Psychological Review*, 85(3), 172.

Moray, N. 1959. Attention in dichotic listening: Affective cues and the influence of instructions. *Quarterly Journal of Experimental Psychology*, 11(1), 56–60.

Newell, A. and H.A. Simon. 1972. *Human Problem Solving* (Vol. 104, No. 9). Englewood Cliffs, NJ: Prentice-hall.

Oxford University. 2021. Oxford English Dictionary. https://www.oed.com/

Pavlov, I.P. and W. Gantt. 1928. Lectures on conditioned reflexes: Twenty-five years of objective study of the higher nervous activity (behaviour) of animals.

Patino, B. 2020. *La civilización de la memoria de pez*. Alianza Editorial.

Pendergrass, W.S. and C.A. Payne. 2018. Danger in your pocket: A case study analysis of evolving issues related to social media use and abuse through smartphones. *Issues in Information Systems*, 19(2), 56–64.

Petersen, S.E. and M.I. Posner. 2012. The attention system of the human brain: 20 Years After. *Annu Rev Neurosci*, 35, 73–89.

Posner, M. I., and S.E. Petersen.1990. The attention system of the human brain. *Annual review of neuroscience,* 13(1), 25-42.

Posner, M.I. and S. Dehaene. 1994. Attentional networks. *Trends in Neurosciences*, 17(2), 75–79.

Przybylski, A.K., K. Murayama, C.R. Dehaan and V. Gladwell. 2013. Motivational, emotional, and behavioral correlates of fear of missing out. *Computers in Human Behavior*, 29(4), 1841–1848.

Raz, A. 2004. Anatomy of attentional networks. *The Anatomical Record Part B: The New Anatomist: An Official Publication of the American Association of Anatomists*, 281(1), 21–36.

Ritzer, G., P. Dean and N. Jurgenson. 2012. The coming of age of the prosumer. *American Behavioral Scientist*, 56(4), 379–398.

Roselló i Mir, J. 1998. *Psicología de la atención: Introducción al estudio del mecanismo atencional.* Pirámide.

Rueda, M.R., J.P. Pozuelos and L.M. Cómbita. 2015. Cognitive neuroscience of attention from brain mechanisms to individual differences in efficiency. *AIMS Neuroscience*, 2(4), 183–202.

Ruiz, A. and S. Cansino. 2005. Neurofisiología de la interacción entre la atención y la memoria episódica: Revisión de estudios en modalidad visual. *Revista de Neurología*, 41(12), 733–743.

Sánchez-Carbonell, X., M. Beranuy, M. Castellana, A. Chamarro and U. Oberst. 2008. *La adicción a Internet y al móvil: ¿ moda o trastorno?* Adicciones, 20(2), 149–159.

Sánchez-Carbonell, X. and U. Oberst. 2015. Las redes sociales en línea no son adictivas. *Aloma: Revista de psicologia, ciències de l'educació i de l'esport*, 33(2), 13–19.

Simon, H. 1969. Designing organizations for an information-rich world. Brookings Institute Lecture. https://tinyurl.com/yb6bllce

Shapiro, C. and R. Varian. 1999. *Information Rules: A Strategic Guide to the Network Economy*. Harvard Business Press.

Skinner, B.F. and R. Ardilla. 1975. *Sobre el conductismo*. Fontanella.

Sohlberg, M.M. and C.A. Mateer. 1989. *Introduction to Cognitive Rehabilitation: Theory and Practice*. Guilford Press.

Sohlberg, M.M. and C.A. Mateer. 1987. Effectiveness of an attention-training program. *Journal of Clinical and Experimental Neuropsychology*, 9(2), 117–130.

Sokolov, E.N. 1963. Higher nervous functions: The orienting reflex. *Annual Review of Physiology*, 25(1), 545–580.

Treisman, A.M. 1964. Selective attention in man. *British Medical Bulletin*, 20(1), 12–16.

Throuvala, M.A., M.D. Griffiths, M. Rennoldson and D.J. Kuss. 2019. A 'Control Model' of social media engagement in adolescence: A grounded theory analysis. *Int. J. Environ. Res. Public Health*, 16(16), 4696.

Tudela, P. 1992. Atención. *In*: J.L. Trespalacios and P. Tudela (Eds.), *Atención y Percepción.* (pp. 119–162). Longman.

Valencia, G. and P. Andrade. 2005. Validez del Youth Self Report para problemas de conducta en niños mexicanos. *International Journal of Clinical and Health Psychology*, 5(3), 499–520.

Van Koningsbruggen, G.M., T. Hartmann and J. Du. 2018. Always on? Explicating impulsive influences on media use. *In*: P. Vorderer, L. Hefner, L. Reinecke, C. Klimmt (Eds.), *Permanently Online, Permanently Connected: Living and Communicating in a POPC World.* (pp. 51–60). Routledge.

Wang, Z. and J.M. Tchernev. 2012. The "Myth" of media multitasking: Reciprocal dynamics of media multitasking, personal needs, and gratifications. *Journal of Communication*, 62(3), 493–513.

Watson, D.J. 1947. Comparative physiological studies on the growth of field crops: I. Variation in net assimilation rate and leaf area between species and varieties, and within and between years. *Annals of Botany*, 11(41), 41–76.

We Are Social. 2021. Digital 2021. https://wearesocial.com/digital-2021

Part II
Predictive Consumption?

CHAPTER

3

Predictive Analytics in Digital Advertising: Are Platforms Really Interested in Accuracy?

Oscar Coromina
Malmö University, Bassänggatan 2, Malmö, 211 19 (Sweden)

Introduction

The Social Dilemma (Orlowski 2020) is a short documentary film produced by Netflix about the dangers and risks of the increasing central role of digital platforms in our everyday life. The film reflects on how social media users are not in fact the customer but the product and how each small interaction becomes valuable information that helps their almighty algorithms learn how to manipulate us. This process is fictionalized using human actors personified as algorithms that are able to predict when, where and how certain content has to be shown in order to nudge a helpless user into a specific action. Quoting the most famous line from another film, we see how these algorithms are making this user "an offer he can't refuse" (Coppola 1972). Although the Social Dilemma is presented as a criticism and a plea to change things in the social media industry, in the eyes of an advertiser this scene presents social media as an infallible tool to promote products and services. In this chapter we propose to look into the predictive features of digital platforms for marketing and advertising, namely predictive analytics. Predictive analytics comes from applying artificial intelligence techniques such as machine learning to analyse data and predict probable outcomes. One of these predictions are reports that give estimations, forecasts or actionable insights that advertising and marketing practitioners use to inform their decisions. An advanced feature of predictive analytics is the automation of several processes that not many years

Email: oscar.coromina@mau.se

ago were considered too strategic or creative for computers: ad creation, media planning and media buying and results assessment (Li 2019). We investigate which processes they intervene in, how reliable they are, and to what extent they have been adopted by marketing practitioners as well as how well known are their techniques and processes that provide these predictive features. These and other questions related to predictive analytics in digital marketing were addressed in a series of in-depth interviews with 12 practitioners representing different areas of expertise such as media planning, web analytics or, search engine optimization. Unexpectedly, despite the fact that digital platforms provide sophisticated tools to make data available, automate processes and forecast performance, these tools are not seen to be as precise as they are depicted in the Social Dilemma. On the contrary, even though they acknowledge that they are increasingly accurate in their predictions and can outperform humans in many areas for managing ad campaigns, professionals also identify several limitations on the tools and techniques that digital platforms make available to them, they put into question their precision and are required to step in to make their marketing campaigns work as expected.

Datafication and Digital Marketing

As stated in the Social Dilemma, social media platforms register each one of the engagements taken by their billions of users. All the likes, reactions, comments, clicks, purchases, shares and other actions are captured and added into gigantic databases. And not only are users' actions; meta-data from their locations, devices, demographics are also collected and used to enable basic operations of platforms such as profiling users, content distribution and, of course, placing personalized ads. Capturing, analyzing and monitoring people's data to understand – and nudge – human activities is the cornerstone of the datafication process that is changing both the corporate world and scientific research increasing predictive capabilities of information systems (Mayer-Schönberger and Cukier 2013). Datafication is about scale and exhaustivity. Indeed, the large critical mass of users and the granular nature of the data captured results in vast amount of information that can be processed and analyzed to extract valuable knowledge. Because of the large quantity of data available to scrutiny we often refer to it as 'big data', a term that was initially coined to distinguish data sets large enough to require computers with a higher level of performance than the ones designed for personal use (Manovich 2011). Big data analysis has become popular during the last decade due to the potential benefits of relying on large scale analysis to find patterns and relations that otherwise would be elusive to inquiry (Boyd and Crawford 2012). As Van Dijck (2014) points out, big data has become a new paradigm in our society and companies and governments look at it as a new valuable stock such as gold and oil, the term data mining itself can be understood as a gold rush metaphor, and be taken for granted that it is both useful and legit. The same reasoning has extended to a science that has found in social media data a superior method for sampling, where N equals the whole population. Researchers now have access to what people do instead of what people say and are thus less intrusive. The adoption of the new paradigm has been faster than it should be and this has left no room for questioning if the trust and belief we put into data analysis is grounded on scientific soundness and is ethically acceptable (Dijck 2014). The first, because bigger data does not necessarily mean better data: it encompasses a bigger effort in separating the wheat from the shaft, increases the risk of mistaking the noise with the signal and does not remove the need for a posterior interpretation that must rely on context and critical assessment (Boyd and Crawford 2012). The second, because the ultimate goal of predicting and changing behaviour shared by governments, companies and academics can come at the expense of privacy and citizen rights (Couldry and Yu 2018).

These discussions and reflections about ontological, ethical and epistemological issues around datafication and predictions built on big data analysis are rarely present in marketing scholarship. On the contrary, they have focused on the sunny side and embraced the promises of more sales and improved brand awareness at lower costs (Dwivedi et al. 2021). Accordingly, the focus has been in developing frameworks for embedding digital analytics in firms practices (Gupta et al. 2020) and studying how specific practices around marketing and advertising are experiencing profound transformations because of digital analytics (Nosrati et al. 2013). This literature renders visible how datafication has changed the ways in which firms perform their marketing operations, manage budgets and marketing plans, deliver personalized products and services and what are the legal issues that enter into play when users data are captured and analysed (Wedel and Kannan 2016). It would not be fair to say that marketing scholars do not address the negative aspects of digital marketing. They definitely apply these methods but they are more interested in the consequences of negative word-of-mouth by social media (Balaji et al. 2016), on how digital marketing techniques such as personalized advertising and retargeting are perceived as excessively intrusive (Truong and Simmons 2010), how the lack of expertise in a digital advertisement can backfire and result in disappointing results (Aswani et al. 2018), the impact of ad blocking, digital fraud and how measurement features like estimations can mislead practitioners and add uncertainty in their marketing plans (Gordon et al. 2021). The present investigation wants to provide a new point of observation to the consequences of the datafication process and staying in the middle ground between the critical approaches to the datafication process and marketing scholars interested in the application of predictive capabilities of algorithmic techniques into the practices of digital marketing and advertising professionals. Digital advertising or digital marketing are broad concepts that are often used indistinctively and may include various forms of promoting products and services such as Internet advertising, Virtual Reality, Augmented Reality, Artificial Intelligence, Robotics, Internet of Things and others (H. Lee and Cho 2020, Grewal et al. 2020). In this chapter this term is also used with a broad meaning to designate the main activity of using digital platforms to promote products and services.

We focus on one particular facet of digital marketing and digital advertising: predictive analytics. Predictive analytics comes from applying artificial intelligence techniques such as machine learning to analyse data and make reliable predictions about human behaviour and/or the performance of a website or app (Mackenzie 2015). This comes in various forms: from reports, forecasts or actionable insights that advertising and marketing practitioners use to inform their decisions, to the automation of several processes that take strategic decisions such as deciding budgets and creativities, target audiences. In between, these predictions in real time can also be used as feedback to adapt or change the initial strategy being used (Li 2019), a process also known as optimization. Even though prediction is a concept commonly related to making accurate forecasts about the future, in this piece we also consider it as a prediction of the process of shedding light on unknown facts or events that may have happened in the past but are needed for algorithmic and machine learning techniques to be better understood. With this consideration we define predictive analytics as the specific processes for finding patterns and shaping data to predict categories, rankings, numbers, groups and labels. These include techniques like neural networks, naive Bayes classification, semantic analysis, linear regression models, decision trees and others that have been used for many years in computer science or very specific business domain like risk assessment (Mackenzie 2015, Rieder 2020) and now are embedded in the most common and widely adopted tools and services by digital marketing professionals.

Undoubtedly, the vast amount of data made available by the datafication process combined with sophisticated predictive techniques have the potential to impact positively in the income statement and that explains how fast the companies have adopted the new paradigm (Bradlow et al. 2017). But predictive analytics also poses new challenges for these firms: the need to adapt technical infrastructures, employee capabilities, managerial models and acquire tools for data analysis (Ghasemaghaei 2019). This can represent an obstacle for small/medium sized enterprises that are in danger of being left behind (Coleman et al. 2016) because of a new digital divide. In the particular area of digital marketing, digital platforms have made sure that this does not happen and provides, habitually for free, tools as well as knowledge for advertisers and digital marketing professionals. This includes tutorials, guides, help centres, dedicated websites, periodic reports, forecasts, APIs, and tools that provide data, information and features that assist advertisers and digital marketing professionals in doing market research and planning. And it has worked out. Almost 60% of the most visited websites are using traffic analysis tools such as Google Analytics or Facebook Pixel and about 45% of them are affiliated with an advertising network like Google ads or Facebook ads (Usage Statistics and Market Share of Advertising Networks for Websites, June 2021 n.d.). Traffic analysis tools and specially advertising management tools have been crucial for the democratization of predictive analytics and to a greater or lesser extent all the routines of the marketers and advertisers rely nowadays in the predictive capabilities made available by digital platforms.

This chapter points to three primary functions of predictive analytics. First, we consider all the estimations, forecasts and information that digital platforms offer to assist digital marketing professionals in planning campaigns and more specifically for defining the target, the budget and the expected results of advertising campaigns. Secondly, the features that allow different degrees of automation for managing advertising campaigns in real time. And finally, the information provided by web analytics tools about the performance of digital advertising campaigns.

Digital Platforms as Advertising Companies

Advertising is nowadays the most important income source for digital platforms such as Google, Facebook, YouTube or Twitter. Like television and other traditional media such as radio and newspapers, these platforms sustain their business model in their ability to generate large audiences and sell them as a commodity to advertisers (M. Lee 2011, Smythe 1977). It is undeniable that these companies are able to attract vast amounts of users. According to Alexa, Google is the most visited website in the world followed by YouTube (Alexa – Top Sites n.d.) with more than a billion active users a month each (Alphabet Inc 2020). Facebook accrues more than 2.8 billion monthly users that access it either via the web or using the app (Facebook Inc 2021). And if we pay attention to the advertising revenue data from these companies, it is crystal clear that digital platforms are also extremely successful in selling advertising space. In 2020, Alphabet, which owns Google and YouTube, reported a yearly total income of more than 168 billion USD, around 85% of which came from advertising (Alphabet 2021). In the same year, Facebook (and Instagram) accounted for a revenue of almost 86 billion, 84 of which came from advertising (Facebook Inc 2021). If we continue to assimilate digital platforms to media, we can say that in both media families content is used to lure the audience. But the newest take advantage of their technological affordances in order to distribute in a more efficient way than traditional media. That is the role of content personalization systems that supposedly are able to predict what is relevant and what is not based on previous interactions and profiling techniques. Thanks to that, these algorithms have become the new gatekeepers in our society (Bozdag 2013, Mittelstadt 2016) when it comes to the circulation of news. And the same degree of personalization is used to deliver ads to a specific target audience. Like with content, digital platforms also rely on algorithmic methods that take into account behavioural data to predict what is the best destination for a given ad of a given product (Keyzer et al. 2015, Nosrati et al. 2013, Winter et al. 2021). Personalized advertising placement and a higher degree of accountability represent the most visible competitive advantages of digital platforms advertising when compared to advertising in traditional media.

If the user is the product and advertising is the business, then we can say that advertisers are the actual customers of companies like Google or Facebook. And to them digital platforms offer an additional advantage: backend systems created to design, implement, analyse, manage and buy advertising space. These systems, on the one hand, make it available in an easy to use graphic interface with sophisticated features based on artificial intelligence to place ads and monitor and optimize performance. On the other hand, they have the potential to replace intermediary figures such as the account manager (M. Lee 2011). However, this has not been the case yet. On the contrary new intermediary figures such as digital advertising/marketing agencies are providing the expertise and know–how, necessary to exploit the possibilities of digital platforms for advertisers.

Methodology

To understand how the predictive features of digital advertising tools are being used and perceived by professionals we conducted a series of 12 semi-structured interviews with digital marketing/advertising practitioners. These professionals had different levels of experience, the most seasoned of which had more than 20 years of experience in digital marketing and the least, 5 years. They also had, different specializations such as SEO, Web analysis or media-planners. However, all of them were familiar with digital media planning and web analytics. In that way, we were able to cover the three previously identified primary functions of predictive analytics in digital advertising. All of those interviewed were working at the time in digital marketing agencies (or as freelancers), providing services to other companies and acting as intermediaries between the digital platforms and the organizations advertising their products and services.

Except for a couple of interviews that were done face to face, all of the rest were conducted virtually due to the Covid 19 crisis.

Tools for Prediction

Early in the interviews Facebook Ads and Google Ads emerged as the most popular systems for digital advertising and Google Analytics for tracking user behaviour. These platform tools are embedded in their everyday professional practices and references to particular features and functionalities which were very common along the interviews. For this reason, we have decided to include a short description of the main features of these services to give some context to the reader less familiar with digital marketing. Other tools and services such as Facebook analytics, Twitter Ads, Google Trends, Google Search Console or Linkedin ads were also mentioned but since they were not as central in the shared discourse they have not been described.

Google Ads

Google Ads is the commercial name of Alphabet's marketplace for selling advertising space across different websites and phone apps. Advertising placement is articulated through the service, offering marketers access to the so-called Google Network that includes Alphabet owned sites, partner websites and other advertising supports like mobile phone apps. Google Network is divided into two different groups: 1) Google Search network, which includes Google, Google Maps, Google Shopping and other "search" sites that partner with Google to show ads and 2) Google Display network, which includes Google own social media sites such as YT or Blogger plus "thousands of websites" that are also part of the network. Attention in Google Ads is sold in the form of clicks (both in Search and Display networks), and views (only on the Display network).

Google Ads in the search network have the distinctive feature that advertising placement depends dynamically on how queries are performed and how users interact with the results. Hence, advertisers "only" pay for the users that actually click on the ads. Google Ads assigns a specific price for different "keywords", the queries introduced in the search box when users look for something. The position of the ad is decided on an auction-based model where top positions are awarded to the advertisers willing to pay more than their competitors. Actually, the auction system is centred not around the highest bidder but on the so-called 'Quality Score', a metric that rates the quality of the ad. In the Google Ads system quality is defined by three factors: Click Through Rate (CTR), Ad Relevance, and Landing Page Experience. CTR reflects how likely it is that the ad will be clicked by users, Ad Relevance reflects how relevant the text of the ad is to the keywords used by the user in the search box, and similarly the Landing Page experience reflects how relevant, well organized and useful the landing Page is. Therefore, the Google Ads' Quality Score drives advertisers to focus the design of their campaigns on coherence between keywords, ads and landing pages. In the Google display network target audiences may be defined by demographics, interests or by user behaviour such as recent purchases or previously visited websites.

Facebook Ad Manager

Facebook Ad manager is the tool that allows it to create, place and manage ads in Facebook and Instagram both in their web or app versions. The tools drive practitioners to plan their campaigns based on different kinds of goals: Awareness (creating general interest in specific brands, products and services), Consideration (make people search for more information around advertised products) and Conversion (selling or using products). Facebook defines and reaches the target audience base in location, demographics, interests, behaviour and connections (Facebook users following specific Facebook Pages). Interestingly, once the core elements of the audience are set, Facebook offers additional targeting tools that take advantage of the affordances of the platform such as delivering ads to people with similar interests and profiles, as current customers. Like Google, Facebook ads sell views and clicks, in the latter a bidding mechanism sets the price depending on how many advertisers are trying to reach a specific target. In Facebook ads the equivalent of Google's Quality Score is the Ad Relevance Score that calculates the expected feedback an ad may receive from the target audience. Having a higher Ad Relevance Score can lower the cost of reaching an audience and is essential to optimize the performance of ongoing campaigns.

Google Analytics

Google Analytics is a web analytics tool from Google launched in 2005 after the acquisition of Urchin by Google. The tool captures, measures and reports website traffic and their main features include information about users' profiles, tracing the origin of the visitors to the website, analysing user's behaviour and tracking conversion (when a specific goal like selling a product is completed in a specific website). In order to use Google Analytics, the website needs to embed a set of JavaScript tags in the source code and this can only be done by the owners of the site. Thus, Google Analytics is essentially a tool for internal use that is unable to give information about competitors. One of the main features of the tool is its capacity to attribute conversions to specific channels to reflect which website, or advertising platform, deserves the credit for a sale or other predetermined goals.

Prediction for Planning

Normally, media planners start from a budget previously established by their client, in this case the company that advertises the product. This budget, alongside with the duration, the maximum spend by day and the target audience are introduced into the planning tools offered by Google Ads and Facebook Ads, which from here offer a first estimate for different indicators of the campaign. Features and estimations are quite different from one platform to another and this reflects the different ways in which digital platforms allow to reach to a target audience. In the case of Google Search, media planners start by defining the keywords that trigger the ads. In the case of Facebook, target audiences may initially be defined by interests and demographics. On top of that both platforms allow additional segmentation options such as location, devices, etc. Once these preferences are set, both platforms provide numbers that account for the visibility of the ads and the price for that visibility. Interestingly, those numbers are not binding because the final price is set after a bid between all different advertisers competing for the same target and when campaigns start to run these estimations become rapidly obsolete and more often than not differ from what was predicted. "It's like using your finger like a weathervane", describes Patrick, media planner, "but when you start from scratch with a new client or a new product these estimations come really handy. However, when you already have been working with a specific product for a while I tend to draw from historical data from other campaigns". Not all digital platforms are equally reliable or precise. All the media planners agree on the fact that Google Ads is more trustworthy and is richer in features than Facebook Ads. "I actually don't consider Facebook Ads a planning tool; you simply introduce your budget there and get an estimation of what you can achieve with that money. While Google does the opposite, it tells how much money and where you have to spend it to achieve results", explains Jordi, media planner. "The real problem with Facebook or Linkedin is the fact that cost estimations are based on intervals, and a range of 2–6 Euros per click doesn't help that much when you are looking for thousands of clicks", explains Cristina.

In our interviews practitioners do not consider this discrepancy between what was predicted and the actual outcome of the campaign a big issue, and highlight that planning tools may not be precise, but are useful while they are taken "with a grain of salt", said Luciana, performance director of a medium sized interactive agency. It is a common assumption that this lack of accuracy is because of the dynamic nature of digital advertising "everything, the price, the number of views, the results, is uncertain and depends on what your competitors do and how well they do it. That is why the plans you initially get from the platforms are just useful until the moment you activate the campaign", explains Luciana. Antoine, a freelance SEO specialist finds an additional explanation: "At the end of the day these tools for planning play an important role in luring media planners to buy ads on the platform, and promising cheaper prices and optimistic estimations about the reach and results of a campaign is part of the platforms marketing strategy. While it is true that there are many dynamic factors that can alter their predictions, I believe that they could do a much better job and present more realistic budgets". The bright promises emerging from these estimations prevent media planners from using these predictions with their clients. "I learned the hard way that you should never show plans made with these tools to your customers. Very soon are going to find that you deliver less and for a higher price from what was forecasted, and they will blame you for that", explains Antoine. However, some companies demand access to this information for marketing agencies and the media planner needs to "be very pedagogic about the plans and explain that the numbers are just approximations. And tell them that they must wait until the campaign starts to have real data to evaluate your planification", confirms Luciana. With the experience acquired by years of planning campaigns some practitioners prefer to tweak these numbers and present their own predictions for the campaigns. "I basically correct upward the estimations that involve pricing, I rely on data from previous campaigns and if I don't have it, I make an educated guess", admits Cristina.

Prediction for Automation

Once the ads are running, media planners have access to different automation features. On the lowest level of automation campaign managers receive suggestions to increase the reach and effectiveness of their initial planification. On the highest level, platforms are allowed to "take full control" of the campaign within a limited budget. But there is a lot of automation that takes place without any intervention from the media planner. "When a campaign is set, you are asked to include different creativities, versions of the same ad. After some testing the platform is able to find out which of the ads are going to perform better and pushes the investment to the most effective creativities", explains Jordi. "It functions incredibly well, it certainly demands more work when you create the campaign because you have to design and create alternative texts and designs for several ads, but it is totally worth it", confirms David M., former media planner and copy writer. In the same line Patrick considers that nowadays an important part of his work as media planner is "to feed the algorithms with choices and then let them take the decisions". Besides these features that are by default activated for ad placement, digital platforms are increasingly assuming new tasks and there is a consensus that algorithms already outperform humans in certain areas. One of these such areas is "smart bidding" which consists of setting the kind of goal the campaign is pursuing (visibility, clicks, sales) and then let the machine learning techniques do their work and all the parameters are adjusted towards this goal. "They are outstanding when it comes to managing the bidding process and deciding when to bid and the amount. We started to rely on automated bidding three years ago and the results are absolutely spectacular. Today it is simply not an option to leave this in human hands", says Magali, owner of a media planning agency with more than 20 years of experience in digital marketing. "The smart biding features from Facebook and Google are both excellent. We have been testing a lot with Facebook lately and the acquisition costs have been halved. We don't even have to think of designing specific creativities for different targets like we used to. We upload 10 or 15 different ads to the platform and the algorithm is able to find which is the best ad for each target. It's hallucinating", agrees Luciana. Cristina is particularly happy with the improvements in this area, "that was the most tedious part of my job as campaign manager, it demanded constant attention and you were making decisions without enough information".

Another feature that media planners find especially valuable is the capacity to find similar individuals to those that have proven profitable in your campaign. "We usually start with a quite broad target definition, but at the moment we have some results we ask Facebook to push our ads to similar targets as the ones who have converted or delivered best results. This works extremely fine", explains David M. Effectiveness in this case comes at the expense of transparency loss: "you don't know the criteria in which similarity is built, may be age, location or whatever but the results are remarkable", recognizes Magali. Although better results are normally good enough in digital marketing, several of our informants are convinced that more information and intervention from the human experts could deliver superior results. "In my experience optimization works better when algorithms and humans collaborate. I am sure that more transparency on the platform side could help me in improving my results", suggests Antoine. I have also found that this opacity is also preventing some media planners from relying on this automation features. "It may seem conservative, but I don't want to lose control of which websites are allowed to show my ads. And when you go for "smart" options you can't decide that. And even after the campaign is finished you don't have access to this information", complains Cristina.

There is also the suspicion that platforms may in certain occasions push the campaigns towards a higher expense. "Platforms are judge and part of this business, it is clear that they want you to spend as much money as possible and they build their algorithms with the same purpose" reasons David B., web analyst. This is rendered visible, for example, when Google Ads suggests additional keywords to increase the reach of the campaigns: "I would never accept by default any of the suggestions for new keywords. They are way too broad and I know from experience that this would only lead to finishing my budget faster without reaching audiences with buying intention", agrees Cristina. And it does not seem to be because this is too difficult for the predictive techniques of digital platforms. "They are capable of doing more sophisticated things and they have access to all the information they need. I trust in the algorithms. But I do not trust in Google", summarizes Antoine. On the other hand, there are also practitioners that tell a different story. "With features like the performance planner form Google Ads we have been able to reduce our costs significantly", reveals Magali. The performance planner is one of the latest additions to Google Ads and draws from simulations and machine learning techniques to deliver "fine-tune" forecasts (Google n.d.(a)). "Differently from the other planning tools, this relies on historical data from other campaigns, and it is much more consistent, especially for campaigns with a higher budget", reveals Magali.

According to my experts, algorithms also struggle to do their job when they are asked to manage campaigns with smaller budgets. "Predictive analytics needs scale to work properly, and this means that only clients with important budgets can fully take advantage of their affordances", recognizes David B. For small-medium sized companies professionals agree that automation features are not that useful. "In our company we have a rule of thumb, if our client is not spending more than 100 euros per day in a continuous campaign, we don't use smart bidding", illustrates Luciana. There are some expectations that in the coming future artificial intelligence would be able to perform better with smaller accounts but not all agree on that. "I think that this is not a priority for the platforms, they are in it for the money and the low-hanging fruits are also the most profitable", considers David B.

Fortunately for the employment perspectives of the interviewed algorithms are not equally infallible when it comes to other tasks and humans are still better making strategical and creative decisions. "Today the artificial intelligence of Google is able to create text ads and decide the keywords but they are not as good as my employees. There is still a long road ahead for them", considers Magali. When algorithms act as creative directors they are not exactly imaginative and subtle. "For the sake of curiosity, I keep looking at the proposals made by Google

Ads for text ads. They are well written, straight forward and simple. But they are not attractive or persuasive and often miss important details that are essential for communicating your product". The same could be said about making strategic decisions that is perceived as the last frontier for algorithms, "I can imagine that in three years our tasks as media planners would be reduced to deciding the strategy and letting the artificial intelligence to the rest", says Luciana.

Predicting Results

During the interviews, we observed that even though all platforms offer reports that account for the results obtained in real time, it is a common practice for professionals to rely on additional web analytics tools to verify the results. Surprisingly, although the possibility of measuring and tracing each of the actions of the users is one of the main assets of digital marketing, there is a clear consensus that the data from Google Ads or Facebook Ads must be contrasted. All my informants complement the reports from advertising tools using Google Analytics. "There are a lot of discrepancies between the tools, metrics differ significantly from one platform from another", signals Jordi. And even tools from the same environment show different numbers. "As incredible it may seem, I get different numbers from Google Adwords, Google Analytics and Google Search Console", explains Antoine. With some resignation practitioners say that they must accommodate with this reality and understand indicators as "orders of magnitude".

A more sensitive issue is the so-called attribution models. That is, the way in which these platforms are awarded credit for the accomplished goals. "According to Facebook Ads, all sales are thanks to Facebook," Cristina ironizes. The same could be said about Google, "If you stand by the default options of Google Analytics you are going to get the impression that most of the conversions come from Google", acknowledges David B. In Google Analytics, for example, the default attribution model awards the completion of the goal to the last "touch point" before the purchase (Google n.d.(b)). So, if a user clicks on an ad placed in Google, the sale is going to be attributed to Google Ads. This seems quite straight forward, but complications may appear when the process is more elaborated such as when a user visits a website after clicking on a banner in Facebook and after some hours or maybe days comes back by searching for the website in Google. Since the last touch point was Google, the search engine gets the credit for the sale. "This is an absurd simplification of the purchase process, normally there are many channels involved in a sale. And some of them are not measurable because they are offline", adds Jordi. To obtain more reliable information web analysts and media planners can tweak the default options or rely on the assisted conversion metrics that measures any interaction, other than the last click, that presumably contributed to the sale. Again, suspicions appear on the side of the media planners when they try to find an explanation for the inaccuracy of the default options.

"Attribution is essential to inform decisions on where to invest your marketing budget, logically all clients look at the return of investment and when they see that a certain platform delivers most of the sales, they want to move the budget there", explains Jordi. For this reason, media planners are very cautious also when they present campaign results to their customers. "There is a lot of elaboration on the data provided by the platforms before reporting results and a lot of pedagogy when you explain it to the client", admits Cristina.

Conclusions

In my conversations with digital marketing professionals, I have found that in their professional practices they are in close contact with the predictive features of digital platforms and have a privileged position to account for the affordances and limitations of the different techniques in play. It has become clear that predictive analytics is not science fiction but an everyday tool for digital marketing practitioners. What is still speculative futurism is the portrayal of algorithms as precise and infallible selling devices and human intervention seems still crucial to obtain the best outcome of advertising campaigns. This is particularly true when it comes to creative work and strategic decisions while in managing the bids for the campaign it is also clear that the predictive capabilities of machine learning techniques outperform professionals delivering best results and there is no coming back to manual management. However, the algorithms are data thirsty and need to work from scale to work efficiently and small-medium sized companies have a more difficult access to the advantages of automation. Automation mechanisms are blackboxed and their opacity prevents professionals from participating and collaborating in the optimization of the campaigns, which is seen as a lost opportunity to obtain even better results. If digital marketing is now a distributed accomplishment between humans and algorithms, the role assigned to marketing professionals and users is to feed the algorithms with different choices upon which the software decides. The efficacy of predictive analytics for managing campaigns contrast with inaccuracy of their forecasts for planning campaigns and also for attributing success to specific channels. In both cases there is the suspicion that platforms' interests in selling ad placement prevails over truthfulness and accountability. In the first case because estimations always tend to present lower costs and better results than what later happens. In the second because all platforms claim the credit for the sales even though this comes at the cost of misguiding the advertisers. This put media planners in a difficult position where they are forced to hide or tweak the forecasts and reports made by digital platforms to avoid being questioned for their results. All these observations show how the tensions between the interests of the different parties involved in digital advertising may be an obstacle to increase the effectiveness and adoption of predictive analytics.

References

Alexa—Top sites. n.d.. Retrieved 7 June 2021, from https://www.alexa.com/topsites

Alphabet. 2021. *Alphabet Announces Fourth Quarter and Fiscal Year 2020 Results*. https://abc.xyz/investor/static/pdf/2020Q4_alphabet_earnings_release.pdf

Alphabet Inc. 2020. *Form 10-K for the Fiscal Year Ended December 31, 2019*. https://www.sec.gov/Archives/edgar/data/1652044/000165204420000008/goog10-k2019.htm

Aswani, R., A.K. Kar, P.V. Ilavarasan and Y.K. Dwivedi. 2018. Search engine marketing is not all gold: Insights from Twitter and SEOClerks. *International Journal of Information Management*, 38(1), 107–116. https://doi.org/10.1016/j.ijinfomgt.2017.07.005

Balaji, M.S., K.W. Khong and A.Y.L. Chong 2016. Determinants of negative word-of-mouth communication using social networking sites. *Information & Management*, 53(4), 528–540. https://doi.org/10.1016/j.im.2015.12.002

Boyd, Danah and K. Crawford. 2012. Critical questions for big data. *Information, Communication & Society*, 15(5), 662–679. https://doi.org/10.1080/1369118X.2012.678878

Bozdag, E. 2013. Bias in algorithmic filtering and personalization. *Ethics and Information Technology*, 15(3), 209–227. https://doi.org/10.1007/s10676-013-9321-6

Bradlow, E.T., M. Gangwar, P. Kopalle and S. Voleti. 2017. The role of big data and predictive analytics in retailing. *Journal of Retailing*, 93(1), 79–95. https://doi.org/10.1016/j.jretai.2016.12.004

Coleman, S., R. Göb, G. Manco, A. Pievatolo, X. Tort-Martorell and M.S. Reis. 2016. How can SMEs benefit from big data? Challenges and a path forward. *Quality and Reliability Engineering International*, 32(6), 2151–2164. https://doi.org/10.1002/qre.2008

Coppola, F.F. 1972, September 26. *The Godfather* [Crime, Drama]. Paramount Pictures, Alfran Productions.

Couldry, N. and J. Yu. 2018. Deconstructing datafication's brave new world. *New Media & Society*, 20(12), 4473–4491. https://doi.org/10.1177/1461444818775968

Dijck, J. van. 2014. Datafication, dataism and dataveillance: Big Data between scientific paradigm and ideology. *Surveillance & Society*, 12(2), 197–208. https://doi.org/10.24908/ss.v12i2.4776

Dwivedi, Y.K., E. Ismagilova, D.L. Hughes, J. Carlson, R. Filieri, J. Jacobson, V. Jain, H. Karjaluoto, H. Kefi, A.S. Krishen, V. Kumar, M.M. Rahman, R. Raman, P.A. Rauschnabel, J. Rowley, J. Salo, G.A. Tran and Y. Wang. 2021. Setting the future of digital and social media marketing research: Perspectives and research propositions. *International Journal of Information Management*, 59, 102168. https://doi.org/10.1016/j.ijinfomgt.2020.102168

Facebook Inc. 2021, January 27. *Facebook Reports Fourth Quarter and Full Year 2020 Results*. Facebook Investor Relations. https://investor.fb.com/investor-news/press-release-details/2021/Facebook-Reports-Fourth-Quarter-and-Full-Year-2020-Results/default.aspx

Ghasemaghaei, M. 2019. Does data analytics use improve firm decision making quality? The role of knowledge sharing and data analytics competency. *Decision Support Systems*, 120, 14–24. https://doi.org/10.1016/j.dss.2019.03.004

Google. n.d.-a. *About Performance Planner—Google Ads Help*. Retrieved 26 June 2021, from https://support.google.com/google-ads/answer/9230124?hl=en

Google. n.d.-b. *Overview of Attribution Modeling in MCF – Analytics Help*. Retrieved 26 June 2021, from https://support.google.com/analytics/answer/1662518?hl=en

Gordon, B.R., K. Jerath, Z. Katona, S. Narayanan, J. Shin and K. Wilbur. 2021. Inefficiencies in digital advertising markets. *Journal of Marketing*, 85(1), 7–25. https://doi.org/10.1177/0022242920913236

Gupta, S., A. Leszkiewicz, V. Kumar, T. Bijmolt and D. Potapov. 2020. Digital analytics: Modeling for insights and new methods. *Journal of Interactive Marketing*, 51, 26–43. https://doi.org/10.1016/j.intmar.2020.04.003

Keyzer, F.D., N. Dens and P.D. Pelsmacker. 2015. Is this for me? How consumers respond to personalized advertising on social network sites. *Journal of Interactive Advertising*. https://www.tandfonline.com/doi/abs/10.1080/15252019.2015.1082450

Lee, H. and C.-H. Cho. 2020. Digital advertising: Present and future prospects. *International Journal of Advertising*, 39(3), 332–341. https://doi.org/10.1080/02650487.2019.1642015

Lee, M. 2011. Google ads and the blindspot debate. *Media, Culture & Society*, 33(3), 433–447. https://doi.org/10.1177/0163443710394902

Li, H. 2019. Special section introduction: Artificial intelligence and advertising. *Journal of Advertising*, 48(4), 333–337. https://doi.org/10.1080/00913367.2019.1654947

Mackenzie, A. 2015. The production of prediction: What does machine learning want? *European Journal of Cultural Studies*, 18(4–5), 429–445. https://doi.org/10.1177/1367549415577384

Manovich, L. 2011. Trending: The promises and the challenges of big social data. *Debates in the Digital Humanities*. University of Minnesota Press. https://minnesota.universitypressscholarship.com/view/10.5749/minnesota/9780816677948.001.0001/upso-9780816677948-chapter-47

Mayer-Schönberger, V. and K. Cukier. 2013. *Big Data: A Revolution that Will Transform How We Live, Work, and Think*. Houghton Mifflin Harcourt.

Mittelstadt, B. 2016. Automation, Algorithms, and Politics: Auditing for Transparency in Content Personalization Systems. *International Journal of Communication*, 10(0), 12.

Nosrati, M., R. Karimi, M. Mohammadi and K. Malekian. 2013. Internet marketing or modern advertising! How? Why? *International Journal of Economy, Management and Social Sciences*, 2(3).

Orlowski, J. 2020, September 8. *The Social Dilemma* [Documentary, Drama]. Exposure Labs, Argent Pictures, The Space Program.

Rieder, B. 2020. *Engines of Order*. Amsterdam University Press. https://www.aup.nl/en/book/9789462986190/engines-of-order

Smythe, D.W. 1977. Communications: Blindspot of Western Marxism. *CTheory*, 1(3), 1–27.

Truong, Y. and G. Simmons. 2010. Perceived intrusiveness in digital advertising: Strategic marketing implications. *Journal of Strategic Marketing*, 18(3), 239–256. https://doi.org/10.1080/09652540903511308

Usage Statistics and Market Share of Advertising Networks for Websites, June 2021. (n.d.). Retrieved 9 June 2021, from https://w3techs.com/technologies/overview/advertising

Wedel, M. and P.K. Kannan. 2016. Marketing analytics for data-rich environments. *Journal of Marketing*, 80(6), 97–121. https://doi.org/10.1509/jm.15.0413

Winter, S., E. Maslowska and A.L. Vos. 2021. The effects of trait-based personalization in social media advertising. *Computers in Human Behavior*, 114, 106525. https://doi.org/10.1016/j.chb.2020.106525

CHAPTER

4

Consumption Prediction on Netflix: Audience Tracking Analysis Based on the Recommendation Algorithm in Times of Pandemic

Emiliano Lucas Iglesia Albores

PhD. Candidate. Doctorate Programme in Communication and Journalism, Autonomous University of Barcelona, Barcelona (Spain)

Introduction

For over a decade, with the explosion of the internet of platforms (also linked to the concept of platform capitalism), the consumption of audio-visual content has changed drastically. Movie theatres and user-shared content have been rapidly replaced by on-demand consumption within a space of concentration of low-cost productions for the user. Content platforms, whose model was pioneered by Netflix, but which have now been followed by Amazon Prime, HBO and more recently Disney+, now concentrate upon the users' media consumption and compete second by second for the market of the viewers' attention. It is no secret that to them a second of the user's attention devoted to another platform is a second lost to capture their own interaction data.

In their struggle to capture as much of a users' attention span as possible, the power to predict future consumption is fundamental. To do this, the recommendations made by the platforms must be reliable and ensure user loyalty, that is: here I have what I want, what I like.

Against this backdrop, this chapter aims to approach the description of the operating system of the consumption recommendations that Netflix makes to its users. In the first instance, the text reexamines the concepts of algorithms,

Email: info@eliahd.com

database, and artificial intelligence as these are the cornerstone of the platforms' recommendation system. The second part of the chapter describes the results of a study to monitor the consumption of content on the Netflix platform. The research, conducted with twelve individuals over a period of six months, explores three main variables: the number of content viewed; the degree of accuracy of recommendations based on the user's tastes and interests; and the length of time users spend on the platform. The study was conducted between March and August 2020 and coincided with the implementation of different measures restricting user mobility as a containment remedy against the spread of COVID-19. The effect of the pandemic is a key factor in the results of the study.

The last part of the chapter reflects on the concentration of the power to control individual decisions in the technical, non-human systems of platforms such as Netflix. At the same time, based on the results of the observations made, the chapter draws attention to the loss of critical capacity of users (almost dehumanised) and the lack of active awareness of citizens. In addition, the text exposes the presence of a functional awareness of the user that facilitates the frictionless development of relations and interactions with the technological and with the intangible or imaginary systems that underlie the fabric of the internet universe.

Towards a Readjustment of the Basic Concepts of the Shaping of Intangibles

Algorithms

As Alan Turing (2012) already announced in the 1940s, the notion of a machine as an instrument capable of accurately processing a large amount of data and arriving at an exact result in a matter of seconds is only a matter of memory capacity and the processing power of the machines. An algorithm is directly related to this. An algorithm is a tool used to manage a given command, evaluate its variables and establish a result in the shortest possible time. The ordered calculations that make it possible to perform a given operation and establish a result are integrally linked to *the intangible*. Turing (2012) proposes, in his test, that the imitation of intelligence through the machine can be proven if the machine can deceive the human in its own interactions.

Alan Turing conducted the experiment in which the intelligence of a machine can be tested: if it cannot be identified by a human being when compared to another human being performing the same logical actions, then it can be said that the machine is intelligent. In other words, if we have an artefact and a human being doing the same calculations or giving the same answers and a second human cannot distinguish between which answer belongs to the machine and which to the human, this could lead to the artefact or machine being considered intelligent.

The Turing experiment helps to understand what Cardon (2018) suggests. According to Cardon, the relationship between algorithms is what allows the machine to possess an intrinsic predictive capability, or power. Such a capability cannot be performed by a human at the speed with which an algorithmic system can perform it. The two thoughts are related since what Cardon and Turing say is similar. Both authors note that the processing speed and the acquisition of memory power can be handled by machines (Turing 2012), through algorithms (Cardon 2018).

The magnitude of Turing's proposition, reinterpreted by Cardon, is signified, and portrayed by Llaneza (2019). Llaneza emphasises data and explains how data is given by consumers to platforms and how they use it to predict new consumption or to generate possible new products. The technical capacity for the analysis of today's algorithms possesses a power that Turing only imagined in the 1940s, but even then, he predicted that incipient computers could have a capacity of thought. The algorithms analysed by Llaneza have the power to predict the future, as they can analyse a huge amount of data and establish a logically defined result. An example of this is the predictive operation of the Netflix platform which, by analysing consumer data, makes product suggestions to its users.

Databases

The construction of a database can provide different points of information that, when analysed and managed, generates new data to be considered in more advanced observations. Each analysis that is carried out allows for an aggregated and more in-depth observation than what the individual and original data provides. A database is a set of data that are grouped and broken down into different sub-categories, which provides an interrelationship between them and, thus, generates new value. To exemplify databases, we can look at the home menu of the Netflix platform. On its central screen there are different films or series on offer, divided into a predefined group of categories (drama, comedy, police, most watched, recommended, etc.). This practical presentation, grouped and catalogued, facilitates consumption and speeds up the consumer's selection.

García Alsina (2017) argues that new technologies enabled the gestation of analysis and the acquisition of new data, based on the accumulation and management of large databases. Reinforcing this thought, Caballero and Martín (2015) argue that the management of databases makes possible the development of new data that can be subjected to new analyses. The new data constructed presents a more accurate and precise view of the core operational functioning of databases: a dataset that supports the creation of new specific data through the cross-linking and analysis of data tables.

Ultimately, new technologies made possible the creation of massive data and the use of the large databases in which they are stored. Without the reduction of costs using simple computers applied in a chain and working cooperatively, it would have been unthinkable to access and work on the amount of data we have today. The existence of large computer farms (data centres), as concrete and tangible places of data storage and processing, demonstrates the physical or real sense, represented in large information warehouses, of the concentration of data.

García Alsina (2017) accurately explains the main qualities that mass data possesses and that were made possible by technological progress. The main qualities are three: speed, volume and variety. These qualities are also mentioned by Fernández (2016), who adds the characteristic of veracity, as well as by Valls (2017) and Marr (2016).

The high-speed processing power of all available data in a very short time makes it possible to analyse and generate new data almost in real time. This process reduces the margin of error in processing and provides agility in the reaction with which the results can be taken on board and exploited.

The use of databases to predict results or situations determined by possibilities can be interrupted and altered by the human factor. This factor is often relegated and assumed to be useless in the face of impersonal, global, and omnipresent Big Data. The interweaving of data generates a sense of completeness and superiority in the face of existing possibilities. This is due to the false sense of the total power of Big Data. However, it must not be forgotten that logical construction cannot compete with spontaneous human interaction.

Strong (2018) and Marr (2016) agree that the creation of a data crosswalk can lead to the prediction of a certain behaviour. Both authors assert that there is a growing awareness on the part of users that their personal data has a certain value. In the face of this awareness, Big Data is affected by the growing refusal to give up one's own data as a user. What was apparently an invisible and unimportant issue has gained prominence in recent years and already affects the reliability of the predictive system of Big Data. Strong (2018) claims that people did not care about what was not visible to them and that for this reason data was given away massively. But this has been changing in more recent years and more and more data is valued by consumers, with a consequent detriment to the predictive capability of Big Data.

In addition to Strong (2018) and Marr (2016), Mayer Schonberger and Cukier (2018) argue that if the world is making its predictions and decisions centred on the tools of Big Data, what is left for humans is instinct, risk-taking, accident and error. These characteristics cannot be logical and are thus marginalised from machine understanding.

Data Acquisition

As Bernard Marr (2016) states, Facebook has 2.5 billion monthly users, who deposit their personal data on the company's servers. The platform can accurately identify users and how they interact with each other. Marr argues that the value of the data that Facebook accumulates is so valuable because it is private. Facebook can access the likes, dislikes, opinions, locations, relationships, banking history, vacation history, habitual consumption and political thinking, among many other variables, of its users.

Valls (2017) points out that the possibility of storing, managing and analysing large amounts of data can only be established thanks to Big Data and the new technologies now available. Changes also focus on the new way of dealing with customers, and how they provide data to manage business profits. Companies like Facebook want to be at the forefront of technology and, to take advantage of data analytics, they need to restructure and constantly produce new information management techniques.

As Duran (2019) highlights, companies such as Facebook are specialists when it comes to storing all the data we provide when using their platforms. An example of this is based on Google's memory when it comes to tracking our steps in all our daily activities. Through tools such as Google Maps, Google Earth, Google Voice, Google Street View, Google Translate, Google Gmail, Google Voice, Google Now, Google Flight or YouTube, the company can always collect the data provided using its tools on any of the devices on which we use its services.

In the eagerness to facilitate household tasks, people can provide all data for free and unconsciously. This data is essential for companies that collect this data, as their advertising businesses are based on the marketing and analysis of personal data. Today, any company that wants to grow exponentially must have a dedicated analytics algorithms department, a team that transforms collected and analysed data into potential profits.

Big Data and data analytics, as Duran (2019) explains, are one of the main bases for the personalisation of commercial offers. Businesses know more and more about their users, and this means that the products they offer are personalised. It becomes very difficult to refuse an offer that is perfectly targeted towards our interests. Duran (2019) indicates the business importance of data acquisition and analysis. The orientation of products towards the needs of customers (many of them created) makes consumption increase exponentially, while the cost of sales production drops considerably thanks to the fact that it is the users themselves who give up their personal tastes or current needs.

Zarza and Murphy (2018) state that there are different types of data, with a certain value in relation to the analyses. These values are given by their scope and ability to be represented. The authors explain that the more personal and specific the data is, the more it will be used for better personalisation of the next product offering. Conversely, the more generic the data to be analysed, the more it can be used for the limitation of control groups and the formulation of products that appeal to a large group of people. These two types of data are of major interest when working with the projection of the future which is introduced under the concept of consumption forecasting.

According to Fernández (2016), there are three forms of data acquisition that companies use in the audio-visual universe: the company's website, social media networks and apps. We can add the purchase of data between companies to these forms of acquisition. This method (not yet regulated by law) is one of the most requested methods by companies that are not able to carry out their own data capture, storage and analysis. Fernández (2016) clarifies that the value of Big

Data and data collection is to allow companies to make predictions, search for patterns and generate groups in which the data is used to carry out tailor-made marketing campaigns.

Emphasising the technical part, as proposed by Duran (2019), implies defining the technical part as the enabler of the materialisation of the process of working with data. As an example of this, we can cite, again, Google, a company that has a detailed record of the movements we make in its browser, as well as throughout the day with its applications. Its predictive data is based on different algorithms and an almost infinite and constant capacity to collect Big Data. So, our searches, our visits to apps, and even the composition of our next email on the platform, can be automated, personalised and predefined to provide us with a very close experience. One could even call it pleasurable. The purpose of the union between the technical processes and the more anthropological elements (such as ideas, culture or thought) is to generate a symbiotic relationship between human and machine. Duran (2019) comments, for example, in an experiment carried out with the *likes* received and given by a user and a specific algorithm, it was concluded that with 300 *likes*, the algorithm was able to know more about the user than his or her stable partner.

But, of course, information or data collection is only useful if the information is then analysed, and a direction is defined for its interpretation and value acquisition.

Artificial Intelligence

While the term artificial intelligence (AI) implies a complex ramification of concepts, AI could be defined as the ability of computers to perform tasks and activities for which human intelligence or reasoning is normally required. The concept, proposed by Rouhiainen (2018), implies a complementarity between logical thinking, the progressive work of algorithms, learning and operations on data, and the use of the learning process for decision-making, emulating the human brain functioning.

According to Belda (2019), the discussion on artificial intelligence is surrounded by a complex philosophical spirit. In his proposal dialogues with theorists such as Turing he asks several questions already formulated by Turing or other AI visionaries: Can machines act intelligently? Are natural and artificial intelligence one and the same thing? Can a machine acquire consciousness?

In a conceptual connection with the problem addressed, Boden (2017) proposes that the central problem of artificial intelligence does not lie in the processing of data, and information, functions, since sooner or later the necessary technological development will be achieved to match brain and robotic functioning. For the author, the real conflict lies in the term "consciousness" and, especially, for Boden (2017), as a principle for life to exist, the type of consciousness wielded as necessary is given in the awareness of one's own existence.

According to Boden (2017), there are currently two types of consciousness defined by artificial intelligence specialists. The first focuses on functional consciousness, which determines that a machine could focus its gaze on different things as needed. On the other hand, this type of consciousness also allows it to become aware of one's own existence.

But Boden also addresses a second consciousness that is marginalised in most research conducted. The second consciousness is the phenomenal consciousness, which focuses on the capacity to create, with the imagination. According to Boden, this consciousness is under-researched and under-appreciated in current research (2017).

To approach the scope of AI, it is necessary to delve deeper into the capabilities that technology may offer in the future. The most distant (and worrying for different researchers) is called the Singularity. The concept refers to the possibility of machines, algorithms or robots having the capacity to replicate and improve themselves automatically. Machines begetting machines. The inescapable goal of the singularity, according to Cortina and Serra (2015), is the fusion between machines and humans. The construction of a human consciousness supported by the intangible and tangible technologies that make up artificial intelligence will be an irremediable conclusion from which, for Cortina and Serra, the new condition of post humanity will be generated.

Kaplan (2017) is another author who sets out the different points of view that analyse the future possibilities of the Singularity. Kaplan and Russell (2004) explain the positive or negative ramifications that artificial intelligence may subject humanity to. Both authors present cases describing the possibilities and consequences of the arrival of the Singularity and support their reflections with previous theories, such as that of Ray Kurzweil, who argues that the Singularity is a manifest destiny of humanity. In contrast, proposals such as that made by Francis Fukuyama warn of the possible dangers to which humanity will be subjected if it continues along the path of technology.

Russell (2004) makes a detailed analysis of the capabilities and assumptions that have to appear for artificial intelligence to acquire the necessary conditions to achieve a Singularity in the near future. Russell's postulates certify that the Turing test has not been passed so far and demonstrates, in a tangible way, that machines are not capable of simulating a creativity of their own. But, as previously expressed, the power of AI is tied to the production of computational and technological power.

According to Rodríguez (2018), who recalls Moore's Law, computational power doubles approximately every two years and, at present, that time is clearly shortening. Rodríguez (2018) states that artificial intelligence will be fundamental for the development of a new humanity and proposes a series of advantages and great benefits that a symbiosis between technology and humans will bring in the future. But developers, engineers, technology entrepreneurs, political leaders, civil society, researchers, and theorists must also take great care. Ethics is a vital issue to watch so that the premonitions of Stephen Hawking and Elon Musk, who agree to limit AI expansion and experimentation, do not come to fruition.

Monitoring Netflix Consumption: The Results

In order to empirically exemplify the functioning of the recommendation and prediction processes of current digital content service platforms, a specific research was carried out. Over a period of six months, the audio-visual consumption of a set of twelve individuals on the Netflix platform was monitored every two weeks. The users ranged in age from 25 to 70 years old. The monitoring of consumption coincided randomly with quarantines or other control measures imposed by the governments of the different countries in which the study was carried out. This situation allowed us to have a precise notion of the increase in audio-visual consumption that took place during the periods of home confinement or mobility restrictions that the people under the study had to endure. Those involved in the experiment were in three different countries, so that the values can be compared according to the geographical area of analysis. The sample countries were Spain, Argentina and the United States. The quarantines were applied differently and in different periods, which is why there are some variations of weeks in obtaining similar results or behaviour in consumption, although, as will be seen below, the behavioural trend is similar. Although it is not important for monitoring the place where the visualisation of consumption on the platform takes place, it is worth mentioning it to better understand the interpretations made on the data retrieved and systematised in different tables and graphs.

It is also important to clarify that the individuals who participated in the monitoring have different characteristics as users and *connoisseurs* of the audio-visual universe. Four people in the group are regular consumers of audio-visual content and have a direct relationship with the media and the visual arts. Five people, the majority in the study, are moderate consumers, with completed university studies and active in the labour market. Finally, three individuals are moderate consumers and are retired people.

The monitoring carried out collected information from users through consumption data tables. The tables for the collection of monitoring data on the content consumed represent a 15-day content viewing measurement period. Likewise, the total consumption and averages, also reflected and recorded in summary tables, provide user data for the six-month observation period (between March and August 2020).

Consumption Measured by Number of Products Viewed

The overall average consumption of audio-visual products viewed was 5.2 productions per monitoring period (every two weeks) over the entire period of the monitoring (six months) which, as indicated, was significantly affected by the quarantine conditions imposed after the outbreak of the COVID-19 pandemic. The highest average consumption per user was 7.3 products (user 9) and the lowest average corresponds to the consumption of user 10, with 2.6 productions consumed on average per monitoring period.

As can be seen in Graph 1, the minimum individual consumption in any of the study periods was 1 and the maximum was 10 contents. In the first measurement of the study, user 3 consumed only one product, while users 4 and 9 consumed 10 products in measurement periods 1 and 8, respectively.

The individual analysis of the users studied describes interesting behaviours. In the case of person 1, the upward variation in consumption is triggered from the fourth period onwards, and almost doubles in the last two measurement periods. The variation between initial consumption (4 contents) and final consumption (7 contents) indicates that consumption during the pandemic quarantine intensifies considerably in the last month of monitoring. In fact, their average consumption (4.8 products) is far exceeded by the behaviour in the last period studied. The second user analysis shows a graph with greater oscillations, but with one of the highest consumption averages among all the users studied (6.1 products). In this

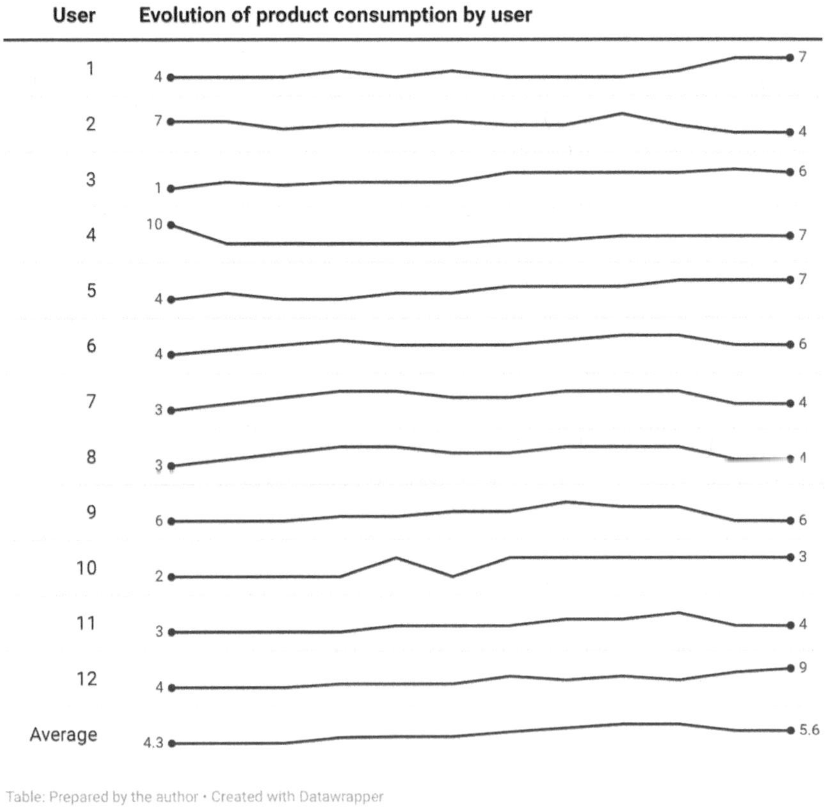

Graph 1: Number of Audio-Visual Products Consumed Per User in Netflix (12 measurement periods)

case, there is a continuous consumption with a deviation of only one point in most of the measurement periods, although there is a peak of visualisation, in the eighth period observed, which coincides with the time of strict confinement. In this case, consumption rises to nine productions, but falls again progressively in the following measurement periods.

Referring to the third user of the monitoring, it is possible to see a significant increase throughout the tables analysed. This is the individual with the highest percentage of variation amongst the entire sample, as he goes from consuming one product in the first period of measurement to viewing seven products in the penultimate period of analysis. User 3 represents the greatest increase in the number of products consumed over the course of the monitoring period. In the first analysis tables, consumption is very low, but from the seventh measurement onwards, consumption rises to six productions, which remains unchanged until the last study period, except for the consumption peak identified in the eleventh measurement period. As in the previous cases, the beginning of the mobilisation restrictions in the pandemic signals a very significant increase in the consumption of content on the platform.

As for user 4, his consumption starts with a very high value, which means his maximum consumption in the analysis period (10 products). Subsequently, content viewing drops to its minimum (five products), which is maintained until the seventh measurement period. From there, a progressive increase begins and continues until the last period studied. During the last four measurements, the number of audio-visual products consumed is 7. The average number of products consumed by user 4 is 6.3, which makes him one of the highest consumers in the sample analysed. User 5, on the other hand, represents a gradual increase in his consumption. The user starts from four contents consumed in the first measurement and increases to five during the fifth and sixth measurements. His consumption increases again to six productions in measurement periods 7, 8, 9 and ends with a final increase (to seven productions) in the last two measurements taken. The difference between their maximum and minimum consumption is low compared to other individuals but shows that consumption increases again in the seventh measurement period, as do most users when they enter the period of quarantine and restrictions for sanitary reasons.

In the case of user 6, a double increase and decrease in consumption can be seen. The user starts with low values (four products) that rise progressively until the fourth measurement period (seven products). Afterwards, consumption drops to 6 products for three consecutive periods, but viewing increases again in the eighth measurement (7 products) and rises to eight products in study periods 9 and 10. Finally, consumption drops again in the last two measurements (six products each). The average of contents consumed remains in the upper range of the table (6.3) and has important values (eight products in the measurements corresponding to the hardest periods of confinement). The decrease in the last two monitoring periods can be correlated with the relaxation of some of the mobility restriction measures.

Users 7 and 8 (who have the same viewing parameter) roughly repeat the behaviour of user 6. Consumption increases progressively until the fifth measurement period (from 3 to 6 products consumed), and then the number of productions viewed drops by one point. From the eighth measurement onwards, consumption rises again by one point, and finally falls by two points (to four content views) in the last two periods of analysis. The data show an average consumption of five products in the period observed and a maximum value of six contents consumed on average per measurement period, one of the lowest in the sample studied.

In the case of user 9, it is possible to conclude from the data of his viewing of audio-visual products that he is an avid consumer of content. User 9 not only has the highest average number of products viewed in the sample during the observation period (7.3), but also has both the highest minimum consumption of all (6) and the highest consumption value in the measurements taken (10). User 9 shows a significant and sustained increase until the eighth measurement period (from 6 to 10 products consumed), but then there is a small decrease (nine products in measurement periods 9 and 10), which becomes deeper in the last two measurements (where it returns to the initial levels of the measurement, with six products consumed on Netflix). According to the consumption data of user 9, the behaviour also describes an increase in consumption between measurement periods 7 and 8, which is established as a behavioural trend among the sample analysed.

The survey data of user 10 in the sample indicate a generalised low consumption behaviour: he is the user with the lowest average consumption in the sample with 2.6 products viewed per monitoring period. In any case, the increase in consumption in the same weeks as the previous users indicates that this user is also affected and influenced by the same factors as the rest of the users, and his increase in consumption in relation to his own data is also relevant, going from 2 to 3 products consumed in the final measurements. The increase in consumption, in percentage, is the least relevant in the sample (150%), but it is still a high increase for the measurement of consumption statistics in the users of a content platform.

User 11, on the other hand, shows a gradual increase since the fourth measurement period. Up to this period, the user consumed three products. From the fourth period onwards, it increases to 4, in the eighth period it rises to 5 and in the tenth period it reaches its maximum with six products consumed. In the last two measurements, his consumption drops to four products viewed. User 11 has an average of four products consumed in each measurement, which places him as one of the lowest consumers in the study. The consumption trajectory, however, also reflects the same trajectory as the previous users.

Finally, the data for user 12 show a significant increase (225%) in the number of products consumed in the period studied. User 12 goes from consuming four products at the beginning of the measurement to consuming nine products at the end of the measurement period, following an upward trajectory. Although measurement periods 8 and 10 reflect drops of one point compared to the immediately preceding one, they are not considered to be significant, as in the following measurements the user exceeds these consumptions and, in the last two measurement periods, the highest values in the time studied are reached.

Details of Aggregate Consumption

The general averages of the sample's consumption shows different relevant trends. The first of these reflects a clear increase in audio-visual consumption on the Netflix platform, which starts in the fourth measurement period and increases clearly and successively in measurements 7, 8 and 9. Likewise, there is a decrease in average consumption in the last two measurements.

The results show that the pandemic encouraged consumption and, after passing the lockdown restrictions, users increased their average consumption compared to the pre-pandemic average. While before the pandemic the average number of products consumed was 4.3, at the highest peak it reached 6.3, and then dropped to 5.6, 235% higher than before the incidence of mobility restrictions in the different countries.

On the other hand, the overall values indicate that 83% of the users increased their content consumption between the first and the last measurement, and only two of them reduced it. Also, at some point in the measurement periods, all users made an increase from their initial consumption, with the only exception of user 4, who had a very high consumption value in the first measurement period. The average percentage increase for all users is 235%, with a mode of 200%, as well as a maximum value of 700% (user 3) and a minimum of 150% (user 10).

Most of the individual graphs of the people analysed (Graph 1) have the same behavioural trajectory, which is evident in the plotted average graph. Except for a few and detailed sporadic high consumptions, all consumptions have an upward trajectory until they reach the peak, and then reduce their level by a few points. However, this end point in the study period is, in most cases, higher than the starting point of the study. The quarantine and the health crisis encouraged a higher consumption of audio-visual products from the Netflix platform among the users analysed, both during and after the confinement.

The Platform's Consumer Suggestions

Having characterised the number of audio-visual products consumed by the study's users on Netflix, the analysis focused on the description of the recommendations made by the platform to users and how they responded to the suggestions made. The suggestions made by Netflix to motivate future consumption demonstrates how the increase in recommendations is also progressive over time, which correlates with the behaviour of the quantity of productions consumed. The correlation demonstrates an important feature of the artificial intelligence process that enables the prediction, discussed in the first section of this chapter. The platform intensifies its suggestions as it learns and processes the user's tastes and consumption using databases, predictive algorithms and the artificial intelligence-guided processes that shape it.

The data recorded in the measurement periods over the study period show a progressive increase in the number of suggestions accepted by users from recommendations made by Netflix. Users have a progressive increase starting at two accepted suggestions, which increases to 2.3 in the fifth measurement period and to 2.7 in the seventh period. The highest points are achieved in measurement periods 8 (3.1), 9 and 10 (3.2). In the last period, users accept on average three suggestions from the platform in each of the analysed time periods (every two weeks). The situation is interesting for analysis and worrying for consumption. If we compare the average number of products consumed by users (5.6) with the average number of products consumed thanks to a recommendation from the platform (3), we can affirm that, on average, Netflix is able to predict or guide 53% of its users' audio-visual consumption. After six months of data analysis, the results indicate that the predictive power of the platform increased by 7% for the users in the sample. Specifically, the power to define consumption increased from 46% to 53% thanks to the increase in the consumption of audio-visual products on Netflix.

The confirmation of the platform's predictive use of algorithms corresponds to two main variables: time elapsed and product selection. Netflix uses these two variables to bring new consumption material closer to the consumer, refining the searches based on the user/customer's pre-selected tastes. The platform's goal is to personalise consumption in detail, thereby establishing consumption captivity and determining the most profitable products.

The minimum value of the average of the suggested consumption among the individuals is 0.2, which corresponds to the user who spends the least time on the platform and consumes the fewest products (user 10). In contrast, the maximum value of the averages among the users is related to one of the highest consumptions in the sample (6.3 products). User 6, with an average of 4.6 contents consumed based on recommendations, consumes up to 73% of his consumption on the platform based on Netflix suggestions.

In any case, as can be seen in Table 4.1, user 6 is not, within the sample, the most recommendation-driven consumer. Almost all the products consumed by user 11 are recommendations made by the platform and, at the same time, 80% of the content consumed by user 7 is also suggested by Netflix. It is worth noting that only users 9, 10 and 12 have low levels of acceptance of suggestions (below 20%). All other users decide more than 40% of their audio-visual consumption based on the platform's recommendations.

The dependence between time spent on the platform and the use of predictive algorithms builds a relationship that is identified in most of the cases analysed. The more time spent feeding the Netflix databases, the higher the accuracy of the algorithms in suggesting new consumption to the platform's customers. This correlation is evident in almost all cases, except for users 9 and 12. The deviations

Table 4.1. Comparison between Average Consumption and Average of Accepted Recommendations

User	Average consumption (number of productions)	Average number of recommendations accepted	Percentage of recommended products consumed
1	4.8	2.25	47%
	6.1	3.08	51%
	4.3	1.92	44%
	6.3	2.83	45%
5	5.5	2.50	45%
	6.3	4.58	73%
	5.0	4.00	
8	5.0	3.00	60%
	7.3	1.42	
	2.6	0.17	6%
	4.0	3.92	98%
	5.8	0.92	16%
Average	5.2	2.5	49%

Prepared by the author

of the correlation in these two cases are explained by the independent consumption demonstrated by the two users in all measurement periods. The data for both in predicting consumption, in contrast to the other users, remains very low: their average acceptance of suggestions is 1.4 and 0.9 respectively, well below the sample average (2.5).

The individual numbers of the analysis tables and the consumption of the analysed users indicates a general increase in the acceptance of the suggestions as the measurement periods approach the end of the study period. This also allows us to verify two positions related to predictive power. Firstly, that the platform improves its prediction over time. Secondly, that the consumer recognises the improvement in the platform's prediction and therefore accepts the suggestion made by Netflix's automated recommendation system more readily.

The data collected in the study demonstrates a relationship between the time and consumer acceptance of suggestions made by the platform as it learns the user's tastes. This behaviour is a central point when analysing the Netflix platform, as predictive algorithms not only establish new consumption for its consumers, but also set parameters for the most watched and profitable products, as well as possible features to produce future audio-visual projects.

With the predictions and with the continuous growth in the acceptance of suggestions by consumers through increased consumption of products, the platform builds an inexhaustible source of information, which reproduces and captivates certainties when it comes to generating and reorganising content. This is a very strong link between consumers and the platform, in which a symbiosis of proposals and the creation of content already approved by predictive algorithms (which learn from the user's consumption) emerges for immediate and future consumption. The risk of generating or contracting products is reduced thanks to the growing exchange of information on consumption between users and the platform.

Returning to the study sample, although the increase in the acceptance of suggestions is constant, it is important to clarify that in most cases there is a small decrease after the tenth measurement period, which is maintained until the end of the observation of consumption. If we take user 6 as an example, we can see how its minimum value is in the first observation period (3) and its maximum value (6) in periods 9 and 10. Then, its values drop to 5 (measurement period 11 and 12) but this decrease is not significant when compared to the upward curve that can be seen in Graph 2.

From the eighth measurement period onwards, a progressive increase in the effectiveness of the predictions is established. The data also determines

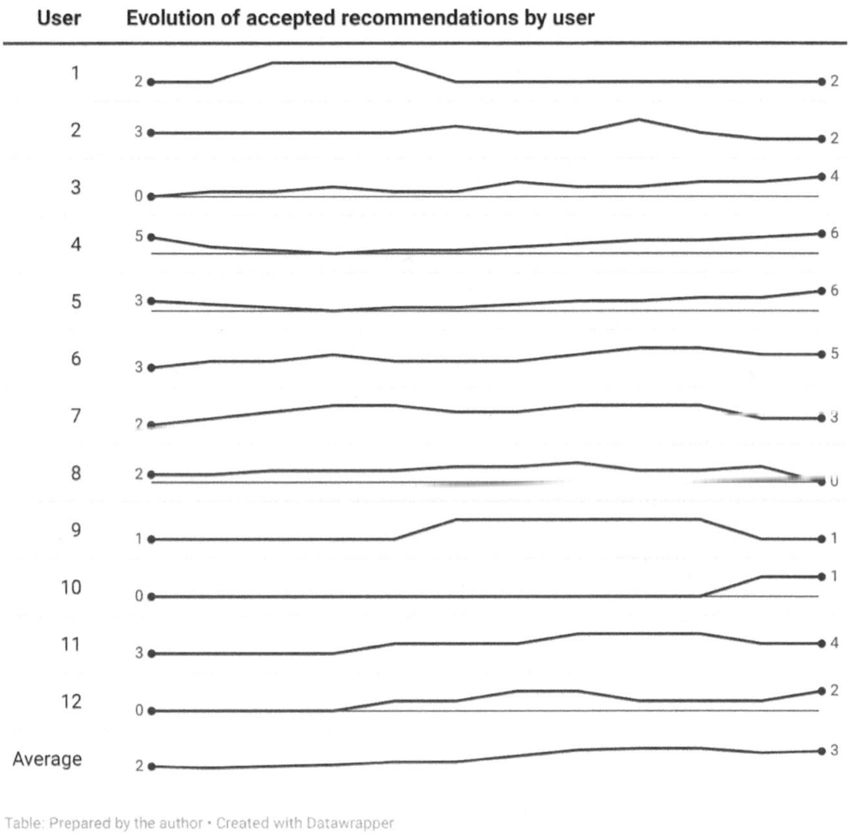

Graph 2: Number of Recommendations Accepted by User

how little time the algorithms need to establish consumer tastes and make more accurate suggestions. The data provided by the user facilitates the complexity and completeness of the algorithm which, with the data received, reaches the necessary parameters for the configuration of user profiles and tastes. A base algorithm is a failed entity if it does not have the complete parameters.

If one looks closely at the different consumption fluctuations and adds contextual data, the results establish a clear parameter with a new focus. During consumption periods 6 and 10, many of the consumers in the sample were in confinement due to the COVID-19 pandemic. During confinement, increased consumer uptake of the suggestions can be observed, which may be a precedent for consumption, with possible explanatory scenarios.

Firstly, Netflix users are more receptive to suggestions as they are deprived of freedom of movement (with more time at home) and use the platform and its content intensively. Secondly, it can be hypothesised that the higher the consumption of the platform due to confinement, the more efficient predictive algorithms are due to the amount of data stored and processed.

As can be clearly seen in Graph 1, the context of the pandemic triggered the consumption of products on the Netflix platform. If we observe Graph 2 and its increase in the acceptance of suggestions by consumers, the values complement each other and correspond to the time factor. The configuration of these three factors creates a perfect dependency between the consumer and the platform. The two actors are symbiotically linked to increase their consumption, to ensure selection and to plan the production of content. On the other hand, the platform ensures user satisfaction when viewing content.

Algorithms are constituted, in this case, from different factors that were altered by the appearance of confinement, a situation that motivated a temporal increase in the speed of actions oriented towards consumption. The time to perfect the prediction was reduced due to the increase in the information received from the users and, consequently, the acceptance of suggestions increased, which, of course, promotes a greater consumption of products.

The Time Factor: Average Netflix Usage

Regarding the time spent on the platform by the users analysed, two points can be made. The first is centred on the observation of the hours that each individual studied and dedicated to the consumption of content on Netflix. For the second measure, it is possible to go deeper into the data collected so far by making a relationship between the different approaches observed. In the first approach to the results, we obtain the average number of hours of consumption distributed in segments: less than one hour a day, 1 to 3 hours a day, 3 to 5 hours a day and more than 5 hours a day.

The study data reveals that most of the users observed spend, in most of the observation periods, between 1 and 3 hours a day consuming Netflix. Out of a total of 144 possible periods (12 users for 12 periods), the first time slot accounts for 49.3% of users' time spent on the platform. The next highest number of time slots found is the one where users spend between 3 and 5 hours a day on Netflix (27.8%). Likewise, 16% of the periods are represented by the band of use of less than one hour a day. The last band, of time, that of more than 5 hours per day of Netflix, only represents 7% of the total study periods of the research.

The data shows that the users consume audio-visual content in the majoritarian time slot, the 1–3 hour segment, with the exception of user 8 who does not consume in this space. Similarly, only individuals 1 and 10 do not consume in the 3 to 5 hour segment. As shown in Graph 3, which counts the number of measurement periods per time slot dedicated to Netflix consumption, the differences between each of the time segments are wide.

Another interesting fact is that almost 50% of consumption time is centred on active viewing of between 1 and 3 hours by all users. This is a significant percentage of hours of consumption that indicates, in at least half of the measurement periods, that users spend between 7 and 21 hours a week on Netflix. This number of hours increases to 21 and 35 hours per week, in the second most representative range (between 3 and 5 hours), which involves 29% of the periods studied.

If we compare the results in Graph 3 with those in Graph 1, we can see the trends. The more time spent on the platform, the higher the consumption of content over the entire period analysed. Furthermore, the global pandemic, together with the compulsory confinement, established a turning point for the

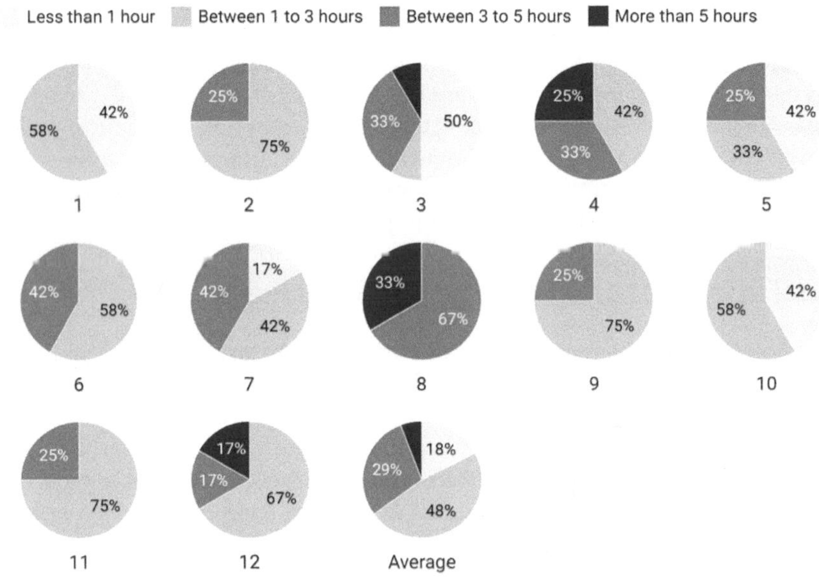

Chart: Prepared by the author • Created with Datawrapper

Graph 3: Time Spent on Netflix by Users and Measurement Periods exponential increase in the consumption of the number of contents and the daily dedication of time spent on the platform.

In this comparison between the two graphs, it can also be seen that consumption time increases as the measurement periods go on. After a start in which screen time was approximately two hours per day, a gradual increase develops until it reaches, on average, 4.5 hours per day. The increase in hours of consumption, in general terms, stabilised at the end of the observations made on the sample (four hours per day). This is a significant increase considering that during the pandemic, audio-visual consumption resources were one of the escape routes. What is interesting is that they do not decrease at the same rate as they increased, and the final values are twice as high as those recorded before the start of the pandemic.

The increase in hours of consumption during confinement reinforces the hypotheses discussed in the previous sections: there is no return to the initial values of audio-visual consumption on Netflix once the 'new normal' has returned to the users studied. At the same time, the more time users spent in confinement, the more hours of use of the platform. Most significantly, the decline in consumption, once the restrictions on people's mobility were removed, was much smaller than the rate of growth over the initial consumption values. On the contrary, they remained at a higher average (over four hours when initially it was about two hours). The individuals, with the symbiotic relationship established with the platform, became more vulnerable to accept the suggestions and to better value its service. The more hours they spent in front of Netflix's content playback systems, the more predictable they became in their behaviour.

Graph 4 clearly shows the progressive growth in consumption time in each of the individuals analysed. The increase is significant and occurs in all respondents. The final values are almost always higher than the initial ones, except in the case of users 8, 9 and 11, although these three cases do have increases in consumption time in the measurement periods 8, 9, 10 and 11. It is interesting to note that the increase in consumption time grows, on average, by 321%, with a minimum of 150% (user 8), and a maximum of 600% (users 2 and 3).

One of the most obvious cases of increased consumption is seen in user 3, as it is the only one that advances through all the defined time frame slots. It starts with the minimum (one hour a day) and increases throughout the pandemic and the development of confinement to more than five hours a day. But, unlike other users, there is no drop-off once home confinement is overcome. His values always increase and do so despite the end of confinement.

As argued throughout the chapter, during the confinement suffered due to COVID-19 (which is affecting humanity and that will continue to do so for a long time to come) the influence of Netflix over its consumers boosted considerably. The increased number of consumption hours and products consumed on the platform generates the permeability that Netflix has as a basis to predict consumption. The growth in usage time generates greater amounts of data that feed its predictive capacity.

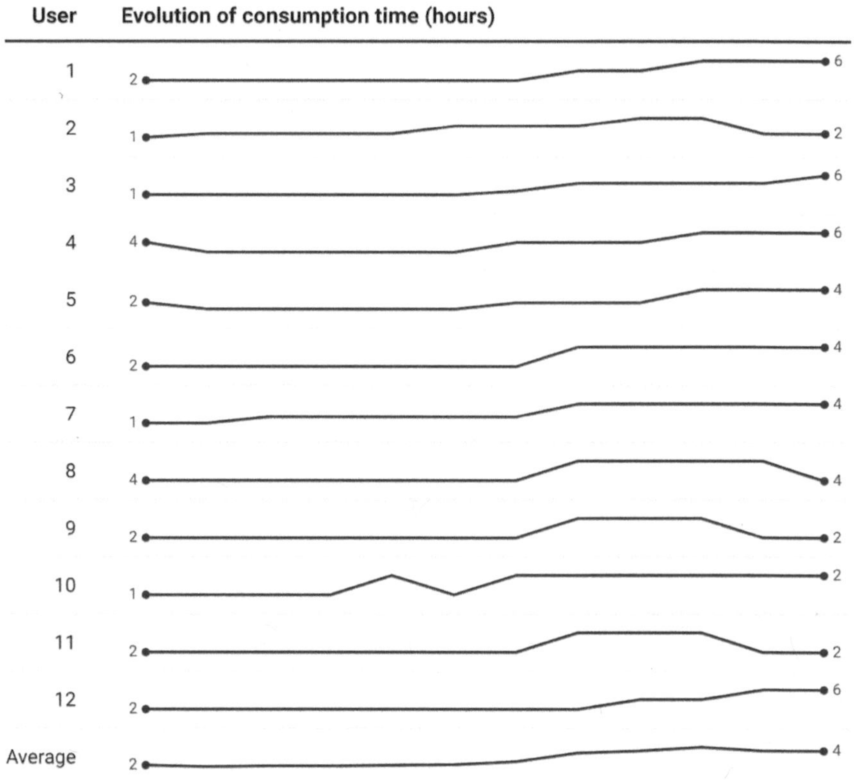

Table: Prepared by the author · Created with Datawrapper

Graph 4: Number of Hours of Content Viewing on Netflix by User and Measurement Period

In the Netflix environment, catapulted by the pandemic effect, even the increase in usage time is being monitored. The platform, which seeks to acquire data, identifies which user is active at any given time. Netflix knows that if the identified user is active, more consumption and interaction data will be obtained from him/her. The strength of the platform and its algorithms is palpable when it comes to data capture, through targeted control of active consumption. To do so, the platform, after a certain period of time, establishes an interactive warning to which the user must respond if he/she wants to resume consumption. This interaction activity is a clear example of the above: the acquisition of active data is fundamental for the economic subsistence of the platform and ensures the construction of its power as a mechanism for accurate (predictive) guidance of users' selections.

The Value of the Human

Taleb (2018) points out that the configuration of a black swan is given by the unexpected, by the feeling of helplessness or inferiority in the face of an event and how humanity must cope with such circumstances. The situation with COVID-19 could be considered a black swan. The pandemic came as a surprise to the world and disrupted the entire daily life of the world. Taleb's (2018) arguments fit in the case of the pandemic suffered by all the countries: powerlessness in the face of the unknown and unfathomable placed human beings on a lower rung of the ladder.

A feeling of fear and mistrust surrounds all decisions that can be taken in uncertain circumstances. For this reason, Taleb (2018) provides an analysis of what needs to be done *a priori*: prevention as a starting point in front of the highly improbable. According to Taleb (2018), prevention plays the role of a primary and unique shield against unexpected circumstances. If the shield falls, everything that follows is an accepted consequence.

Whether something is extremely likely, such as the emergence of COVID-19 or the development of an artificial intelligence with a will of its own (which is entirely probable, given technical advances and the precautions of many experts), it is necessary to be prepared for such a situation, although it may be considered improbable. Even though current technological developments do not usually represent a risk for humanity, critical voices indicate that, in the near future, the film and literary dystopia according to which the value of technology will far outweigh the value of life and may become apparent.

With the exponential and uncontrolled growth of artificial intelligence systems, and given the power of machines, it is logical to think that some major setbacks for humanity may occur with new technological processes and artefacts. Apocalyptic visions of the future are inherently human, a characteristic that machines do not possess. Cold and calculatingly logical, it is machines that will ultimately be able to calculate when selecting a target on the battlefield. It is a warning made by Quirante and Álvarez (2018) when they point out that artificial intelligence may take ultimate control of weapons, to take just one relevant example.

In the case of Netflix, the conjunction of factors (algorithms, databases, artificial intelligence, and the pandemic) translates into a serious problem based on the standardisation or automation of audio-visual and cultural consumption. The constant interaction, seen as a path of user participation, which believes in his/her control over the platform, also reflects the ideological domination exercised by the machine over the users' beliefs. At the same time, it feeds the loop that makes the machine more perfect, which receives more information and thus sophisticates controlled and passive cultural consumption.

The path of prediction limits the possibility of human decision-making and transfers the act of decision to a recommendation, to a subtle, tremendously sophisticated suggestion. In the case of Netflix, it is a decision of audio-visual consumption, but in the range of current uses of the internet, dominated mostly by social media networks, the problem extends to any human decision: regarding relationships or economic, cultural, ideological, or political issues.

The pandemic universe of the second decade of the 20th century does not only describe a black swan in the field of health. The technological scenario, largely dominated and guided by the concentration of data and information, establishes a second black swan that manifests in the unpredictability of the dominance of technology over the capacity of human decision-making. On this path, while users feel more comfortable on the platform, as in the case of most of the users analysed, the decision-making process becomes less and less human.

Having identified the problem of prevention in front of the unpredictable, the black swan, the path opens towards understanding the concept that transcends the whole chapter: the intangible. Human reasoning makes it very difficult to believe in what the eyes cannot see. Artificial intelligence is based entirely on the intangible, hiding the tangible – as a physical implicit – in the concept itself. For most people, thinking of the web, which is known as the cloud, as a set of tangible materials and with very strict physical and market rules, can generate a huge distrust in the concept of the product. Therefore, it is hidden and camouflaged as freedom: freedom of expression, freedom of access, freedom to know, freedom to decide, freedom to move to the left, freedom to discard.

Interaction with the intangible is always more comforting because it has the quality that it cannot be destroyed or corrupted by the human being. Marcuse (1987) argues that where the concept of intangibility generates a false sense of security and hides the true notions of the palpable and its relations to the global market, the perfect system of human domination is mapped out and defined.

Conclusions

The power of platforms such as Netflix is centred on the processing, storage, and interaction of data. These qualities are constantly growing and transform digital platforms into a living entity, with a tangible, constant and impossible to master halo of evolution. The pandemic only accentuated their power. People engaged in a more prolonged and indiscriminate consumption of their content. It is this increase in hours of screen time and number of products consumed that establishes a more vulnerable state in the decisions of consumers. Through the extended hours and products consumed, the platform increases its power as it sophisticates the filtration of its networks of manipulation and control. The more the platform is consumed, the more likely it is that the recommendations made by the platform will be accepted by users.

The data from this research conducted with a small, controlled sample clearly exemplifies the dominance that the platform exercises over consumers' free will. The more a user consumes, the more the content suggestions proposed by the platform are better related with their own preferences and future consumption. The platform's decision-making power increases, while the user's decision-making becomes simpler, more domesticated. In other words, the more one consumes, the more vulnerable one is to the power of prediction.

With only 12 users, the observations carried out systematically for only six months demonstrated the change in consumption trends and in the behaviour of the users analysed. With scarce resources, the data generated by the consumption monitoring show the possibilities of prediction and control that the platforms have with their users. If this data is multiplied by the hundreds of millions of users that Netflix has in the world and is added to the 24-hour-a-day operation, the control paradigm begins to take on its real dimension.

The value of the human is impaired in the face of the intangible power that hides behind the invisible (Foucault, 1979). The emotion of free (uncontrolled) choice generates less profit than the acquisition of new consumption data by an algorithm that never stops proposing consumption. The human is reduced to being another consumption machine that is easily programmable for the acceptance of new products. The more humanity deteriorates, the more useful it becomes to a system of artificially reconstructed moralities and ideologies. In its deterioration, the critical behaviour becomes even less possible.

The path de-ideologized by the intangible paves the way for domination and discourages the critical appropriation of technology. Continuing the uncritical and consumerist path, which accepts without reason the suggestion about the decision (even if it is only for consumption), and brings us closer to the realisation of the formula exposed at the end of this chapter. By the time we would like to realise its dire consequences, it would already be too late to reverse the process of the mechanisation of the human (the loss of meaning).
The final formula:

> *The loss of the value of the human + the systematic consumption of audio-visual products + the transformation from human to machine + the lack of capacity for critical analysis + the rapid advance of new technologies + the lack of legal regulations + the hiding of the tangible behind the intangible + the concentrated global economic power + the political power without moral values + the desperation for the superfluous and constant consumption of products of little affective value: The end of the human being.*

References

Belda, I. 2019. *Artificial Intelligence: From Circuits to Thinking Machines*. RBA Libros S.A.

Boden, M. 2017. *Artificial Intelligence*. Turner Publications S.L.

Caballero, R. and E. Martín. 2015. *The Basics of Big Data*. Editorial Catarata.

Cardon, D. 2018. What algorithms dream of: Our lives in the time of Big Data. Dado Editions.

Cortina Ramos, A. and M.A. Serra Beltrán. 2015. Humans or posthumans? *Technological Singularity and Human Enhancement*. Editorial Fragmenta.

Duran, X. 2019. *The Empire of Data: Big Data, Privacy and the Society of the Future*. PUV.

Fernández, E. 2016. *Big Data: Strategic Axis in the Audiovisual Industry*. Editorial UOC.

Foucault, M. 1979. *Microphysics of Power.* Ediciones de la Piqueta.

García Alsina, M. 2017. *Big Data: Management and Exploitation of Large Volumes of Data*. Editorial UOC.

Kaplan, J. 2017. *Artificial Intelligence: What Everyone Should Know*. Teell Editorial.

Llaneza, P. 2019. *Data Nomics: All the Personal Data You Give Away Without Realising It and Everything Companies Do With It*. Editorial Planeta S.A.

Marcuse, H. 1987. El hombre unidimensional. Ariel.

Marr, B. 2016. *Big Data: Using Big Data, Analytics and Smart Metrics to Make Better Decisions and Increase Performance*. Teell Editorial.

Mayer Schonberger, V. and K. Cukier. 2015. *Big Data: The Massive Data Revolution*. Turner.

Mayer Schonberger, V. and K. Cukier. 2018. *Learning with Big Data*. Turner.

Quirante, R.M. and J.R. Álvares. 2018. *Artificial Intelligence and Autonomous Lethal Weapons: A New Challenge for the United Nations*. Ediciones Trea.

Rodríguez, P. 2018. *Artificial Intelligence: How It Will Change the World and Your Life*. Editorial Grupo Planeta.

Rouhiainen, L. 2018. *Artificial Intelligence, 101 Things You Need to Know Today About Our Future*. Editorial Alienta.

Rusell, S. and P. Norvig. 2004. *Artificial Intelligence: A Modern Approach*. Parson Prentice Hall.

Strong, C. 2018. *Big Data: On a Human Scale*. Editorial Melusina.

Taleb, N.N. 2018. *The Black Swan: The Impact of the Highly Improbable*. Paidós.

Turing, A.M. 2012. *Can a Machine Think?* KRK Editions.

Valls, J. 2017. *Big Data: Trapping the Consumer*. Profit Editorial.

Zarza, G. and J. López Murphy. 2018. *The Engineering of Big Data: How to Work with Data*. Editorial UOC.

CHAPTER

5

Social Complex Networks Analysis as Predictors of Users' Behaviour in the Digital Society

Laura Rojas-De Francisco*, Juan Carlos Monroy Osorio and Santiago Rodríguez Cadavid

EAFIT University, Av Las Vegas # 7-50 Block 26 Office 304, Medellín (Colombia)

Introduction

Every moment of the day users review profiles on social networking sites, to read what is shared with each other, to give feedback by forwarding content, giving a like or another reaction, responding with comments or sharing new content. With this digital use and casual leisure on networks, every user maintains interactions with acquaintances, finds information of interest and fills fragments of time with a simple activity that makes it easy to repeatedly fill the time.

This is useful in several areas of knowledge that study people's behaviours to know or predict a user's actions. In this sense, users are providing valuable information that markets can take advantage of to establish consumption trends. To understand and obtain audience profiles, to send information related to the user's interests. Sometimes, making possible an information bubble, forming what is known as an echo chamber (Cinelli et al. 2021). Consequences are many, such as leading people towards bias and prejudice (Davidson and Farquhar 2020), facilitating manipulation (Förster et al. 2016), or laying a basis to distance themselves from reality (Moqbel and Kock 2018), among many others.

In this scenario, there are initiatives oriented to take advantage of that information from users in marketing, to promote products and services and to reach potential consumers more directly through social networking sites, since

*Corresponding author: lrojas3@eafit.edu.co

these facilitate interactions with and between consumers-users and facilitate knowledge about a user's behaviour and the subjects with which they have interactions for analysis and decision-making (Evans et al. 2021). This has had an effect on how people consume and make exchanges on supply and demand.

What happens between ICT users and the organizations that provide solutions for consumer needs? The use of Social Networks Sites—SNS provides an opportunity to explore consumer behaviour in the communication of shared content between marketers and consumers; users of various digital technologies can provide guidelines for thinking about strategies to influence their decisions (Brodie et al. 2013).

This chapter wants to show how using Complex Social Network Analysis (CSNA) allows us to consider social media as predictors of users' behaviour, because it is possible to explain how people within the digital society build communities online, and to know how by understanding these conversations, marketing makes decisions. This chapter exposes the way to analyse information obtained by downloading large volumes of contents, publicly shared on social networking sites to determine the different relationships and topics shared and commented on by social network users.

This chapter also explains it through examples, showing in graphics the CSNA and giving some descriptions about the meaning of the shape and certain decisions and in a general perspective, what is possible to achieve with CSNA application in consumer behaviour analysis on Social Network Sites (SNS), interactions, comments and sharing activities. The chapter also aims to illustrate one way of recognising the bags of words (Sonawane and Kulkarni 2014) as tools through which users' topics of interest can be identified, as well as recognising, through words, what is shared with others on the network. Using the visualisation of complex social networks through graphs that allow their analytical description, allows the approach to a large mass of data that, when organised visually, facilitates the understanding of the interests shared among users. They form communities of interest, a basic element for decisions in a market.

Analysis of Complex Social Networks in the Production of Content by Consumers-Users

Marketing has tools that have changed the way of buying and consuming products and, in the process, consumer behaviours (Mangold and Faulds 2009) in digital culture, with digital advertising, about effects of digital environments on consumer behaviour, mobile environments, and online word of mouth (WOM) (Stephen 2016). As most consumers as users have embraced information and communication technologies and ICT and produce information each time they need something and consult SNS, that allows them to rethink and formulate strategies while sharing personal information (Thong et al. 2011). Knowing that it is possible to build a relationship in the interaction and communication between organizations and consumers-users (Stephen 2016). In addition, instant access to unlimited information is possible, making conceivable feedback from consumers (Miettinen 2010). The market has an unlimited supply of opinions through SNS. Organizations or marketing agencies can analyse and interpret all the data produced by consumers, when using every moment available to take a look at their profiles and feeds, comment posts or videos, enjoy the GIFs, follow the news, etc. And once a phenomenon is understood, implementing marketing strategies according to those data provided in moments of leisure (Kaplan and Haenlein 2010).

Social media represents a marketing tool for understanding consumer behaviour. By tracking communications and interactions in SNS among consumers, brands, organizations, and other actors, an unlimited supply of contents are likely to be selected, analysed, interpreted, and implemented into marketing strategies (Kaplan and Haenlein 2010). In this scenario, contents are a large amount of data that provide information shared by users, communities and entities, which can be tracked, downloaded by data mining, stored and transformed into structured data, in order to analyse and obtain knowledge about their interests and needs (Hema and Malik 2010). Thus, social networks and microblogs offer raw material to analyse users′ motivations in shared contents, as well as to allow the analysis of multiple short texts and narratives because each new text makes a specific and valuable contribution (Roma and Aloini 2019, Smith et al. 2012).

This is possible thanks to the automation of the data acquisition process and the computational tools generated for the subsequent treatment of this information; it is possible to work on large databases related to many and varied complex networks (Barabási and Crandall 2003). Public access to the enormous amount of data produced in SNS stimulates us to look for information with complex social networks analysis (Boccaletti et al. 2006). In this sense, the discoveries of the small world effect (Watts and Strogatz 1998) and the free-of-scale characteristic of most complex networks are necessary for their understanding. The small world proposed and based on the experiment that led to the "principle of the six degrees of separation" (Watts 2003). Complex scale-free networks are those that assume a degree distribution independent of the scale of the network, it is not homogeneous, these nodes have few connections or a few nodes have many (Aparicio et al. 2015).

Three concepts play a key role in the understanding of complex networks: the mean length of the roads, the clustering coefficient or degree of interconnection, and the distribution of degrees (Furht 2010). Managing the information of the networks individually is difficult. However, Watts and Strogatz' networks characterization allows us to compare similarity and adequacy. With that it is possible to generate graphs showing vertices or nodes joined by links, representing relationships between elements of a set (nodes, vertices, and links), which had a relatively high clustering coefficient, as occurs with regular networks. On the other hand, the scale-free networks based on observing the degree distributions of many real networks, offer a way to measure two quantities and their relationship.

Kozinets's (2012) netnographic approach is ideal for such an SNS exploration, as it allows online observations to collect and analyse public digital conversations in social media and obtain consumer insights (Kane et al. 2014). According to Scott (2011), a social network is extended and is intermingled with social reality in a grid of connections, through which individuals are bound together. Then, in order to understand a social network and the fundamentals for creating or improving engagement strategies, a CSNA is useful to identify and explore interaction patterns and connections between actors (Barabási 2002, Boccaletti et al. 2006). This analysis explores the quality and quantity of the associations and interdependencies among actors because of its explanatory power on social behaviour (McGloin and Kirk 2010), as well as the awareness of network subjects (Crossley et al. 2009) in a flow of transformations, due to the influence of individuals' opinions in the network (Boccaletti et al. 2006).

The outcome of CSNA is a complex network consisting in a graph with nodes and edges that represent relationships or social interactions such as dynamic social systems among actors (Barabási and Crandall 2003). The study of social networks focuses on understanding the topology of networks by analysing nodes that represent dynamical units, and links and paths, indicating users (actors) interactions between nodes; in which communities and cohesive subgroups could emerge (Boccaletti et al. 2006). When the social network can be described as a community, clusters compose the graph: areas within the network that exhibit concentration of connections that determine network membership and boundaries (Millham and Atkin 2018). Visual analysis can be useful in determining network-intrinsic properties, such as Centrality, which shows the importance and involvement of an individual actor or node within a social network. Cohesion is understood as connectedness and togetherness among components of the social network. Or Density, the amount of occurrence of relations and links between at least two actors within a social network (Frey 2018, Furht 2010), which allows us to understand an actor's involvement and engagement to create strategies based on users interest, topics, and discussions.

All the above provide valuable opportunities to explore consumer behaviour in markets by analysing user-generated contents shared on social media. This, in turn, offers open information to support the decision-making processes of managers and generate insights for marketing strategy. However, the analyses of Social Network Sites data comprises characteristics of a complex system (Guliciuc 2014), which require novel approaches to data analysis that embrace complexity.

There are several methodologies that explain how to do certain processes related to these kinds of analyses, such as data mining, graph of words, digital ethnography or thick description for smart and big data as well as tools for monitoring digital conversations like social listening, text spoken interaction and corpus-assisted analysis. Analysing the study methodologies applied to the SNS can be as extensive as are the SNS itself. In CSNA different methods can be used, and this can change the order of the processes required before, during, and after applying CSNA to social media data, but in this chapter our purpose is to show which kind of market insights can emerge from big data visualisation practices. However, the process starts as many types of analysis, by identifying and collecting data, all the way to analysing, interpreting, and presenting a research report. For this purpose, we explain the process of analyses required to understand public information provided by users in SNS, in a strategic approximation to understand online conversations by using data mining, complex social network, graph-based keyword extraction, graph of words, thick description thematic analysis, among other types of analyses, in order to provide the fundamentals of understanding markets.

Methodology for Understanding Online Content for Marketing Strategies

The information we use to share as users in online communities is available and traceable, sometimes following codes marked as hashtag by users or by tracking keywords. Then, it is possible to download this information by data mining or web scraping, techniques and software in which information is extracted from websites having in mind a period and subjects of study. Web scraping is a very imperative technique which is used to generate structured data on the basis of available unstructured data on the web (Saurkar et al. 2018), data mining is a tool to obtain data but also to analyse large, complex, and frequently changing social media data (Saurkar et al. 2018).

The metadata tag known as a hashtag (#tag) used in social networks and microblogging services makes it easier for users to search for messages with a specific topic or content (Aparicio et al. 2015). This also facilitates a way to obtain the same information on different networking sites and collect it in a single flow (SNS) when it is made public. Making it possible for monitoring and analysing and, if necessary, opt for a new data collection.

The process of identifying sources and collecting information is divided into stages that include 1) the collection and storage of information, 2) the treatment of refining information and databases, 3) the data and information visualization and 4) the contents analysis. Each stage is key in data modelling, which results in a piece of structured information to be visualized and interpreted. The data obtained is refined through a data filtering process using analytical software, to do this part it is possible to apply interface protocols, API, designed to obtain and organize historical data of content shared in services such as Twitter, Instagram or Facebook, among others.

The information is transformed into structured data and stored for multiple analysis in a database. Once organized, relationships between subjects and between conversations are searched for by monitoring the SNS. The process makes it possible to extract information made up of large amounts of data to create new quality content (Glez-Peña et al 2013). Then, data can be analysed through a complex network analysis, starting from the messages and identifying relationships in nodes taking into account Scott's graph theory (Scott 2011), in which the nodes in the network are people and groups while the links show relationships or flows between the nodes. A common and standard approach to model text documents is bag-of-words, explained by Sonawane and Kulkarni (2014). The bag of words is a tool that captures the frequency of words and analyses them based on an algorithmic representation. This representation models the relationship and structural information of the texts to characterize them as a graph using two main elements. On the one hand, it analyses the vertex as a characteristic term and, on the other hand, it evaluates the edge relation to determine the link between the characteristic terms, to indicate the weight of each term in the corpus, and to structure the classification that guides the representation and subsequent text analysis.

The modelling applied is carried out taking into account the data mining processes CRISP-DM or Cross-Industry Standard Process for Data Mining (Chapman et al. 2000), which is adapted to each of the cases to understand and analyse the themes found in the data. The data mining process is also useful for obtaining the data and for ensuring its preparation, analysis, evaluation and search for new categories or variables.

Figure 5.1 of the methodological process indicates the most important and frequent dependencies between phases, where the triangles without a line of continuity want to reflect the cyclical nature of data mining itself. In the search for documentary information, actors and entities that talk about an issue that is evidenced in a continuous process after the deployment of a solution means that

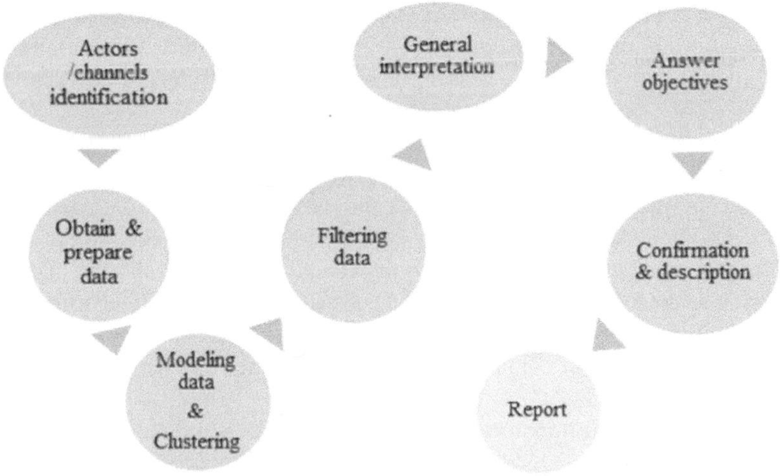

Figure 5.1: Methodology Process
Source: Own elaboration inspired in CRISP-DM Model

the search for data that provides value is an iterative and constant process. Lessons learned in the process can lead to new, often more focused research questions, and to iterative data mining processes which are easier to give information from the already established experience.

The initial phase focuses on understanding the objectives and requirements of the project according to the area of knowledge. The process continues with the definition of the problem and the development of a preliminary plan to achieve the objectives. At this stage, the volumes of information identified are ready to process through the process of scrapping and content scanning. The phase for data understanding begins with an initial collection of data, processes and activities, intending to become familiar with data, by identifying the quality of the problems, discovering first signs of emergent categories or variables and detecting interesting topics or problems to expose hidden information.

The data preparation phase covers all the activities to build the data set. These tasks are executed multiple times and without order but begins by cleaning the texts by removing punctuation marks, other signs, and web addresses. Then, the texts were standardized so that everything is lowercase, accents and other problematic characters disappear (such as emoji) and initial filtering of stop words (common words that do not bring information such as "the", "is", "and"). Tasks include selecting and transforming tables, records, and attributes, and cleaning data for modelling tools.

In modelling, various techniques are selected and applied, and parameters are calibrated to obtain optimal results. If they have specific requirements for the shape of the data, it is necessary to go back to the data preparation phase. The evaluation stage is done on what has been built in quality models that allow a deployment that depends on the requirements, being as simple as the generation of a report or as complex as the implementation of an information exploitation process that goes through the whole situation.

For the identification of nodes, the social network analysis model is through Scott's (2012) graph theory. The nodes in the network are people (users' profiles) and groups (online communities) and links are shown by lines between nodes as relationships or flows.

Data Connection, Relationship Formation, Nodes and Complex Networks

To discover the generic properties of the different types of complex networks accessed, connections are established between users, followers and influencers, or by information distribution patterns. The nodes of the network are glimpsed, by their number of connections and possible groupings by content clustering. That is, by coding the recurring data to the same theme and classifying it into the nodes as thematics. To achieve the characterizations of the information on the network, thematic content analysis is required (Lai and To 2015). That allows developing thematic categories from detecting codes shared by hashtags, keywords, forwarded messages (i.e. Tweets) and conversations in replies (i.e. RT / Re-Tweets). From finding these conversations, both the relationships between the actors and their social actions can be identified and considered as nodes.

For the analysis of the results, the methodology of random networks is used under the Erdõs–Rényi model (Erdõs et al. 2013), which allows us to define a new node which is linked with equal probability with the rest of the network, that is, it has statistical independence with the rest of the nodes of the network. The inputs for this network are conversations (i.e. exchange of texts on the Twitter platform).

For the construction of the network each node represents a word that appears in the texts. An edge with a weight of 1 is formed when two words appear together for the first time in a text. If these already had an edge, then 1 is added to its weight. At the end of the previous process, the network is established, and a division is made into communities. The size of the nodes represents the degree of the node, which is proportional to how many words appear in texts (Sonawane and Kulkarni 2014).

In the modelling of the data collection processes, there may be some gaps since they are not usually constructed to collect data from psychographic and socio-linguistic variables usual in the sociological and demographic study of content and social actors. This process does not obey a construction of a previous procedure to determine the type of data to search because it starts with the collection, storage and analysis of the data that users produce when sharing their messages. The collection and storage procedure, summarized in the following table, shows the collection process and sources used in the process. It also gives the guidelines for the categorization that allows the analysis of the contents.

Table 5.1: List of Technologies, Sources and Categories of Analysis

Collection and storage	Possible information sources	Analysis categories
• Data collection: API Twitter insights • Scrapping flow control API • Non-relational database repository • Sentiments, keywords and identities API	• Social networking sites • Blogs • News in informational portals • Audio and audio-visual channels	• User Profile • User followers • Hashtags t in the content • Date • Retweet count at the time of capture • Replies (conversation) • Sources • Public availability • Language • URL's in the content (shortened) • Attachments (photos, videos, etc.) • Keywords • Geo-referencing (when available) • Word count

Then, the data collection modelling is guided by the achievement of elements associated with the construction of thematic categories (we tested it with volumes of information, while not exceeding 70 thousand contents). The results help to identify and classify data according to topics, words, hashtags, users, some coordinates or communities, and relationships between communities with contexts.

Hashtags collected over a specific period can show related topics. From the search with keywords, social actors can be detected, which are referents to be tracked. Among these, there may be institutional entities or users. However, those who provide more information and identify themselves in the promotion of activities, events and brands are the influencers, on which the same process explained above can be continued to determine, in their behaviour of using the network, those aspects that are decisive for the formulation of a strategy.

After cleansing and analysing the data visualizations (the social network and its clusters), the corpus is ready to provide a deeper understanding of the social media content by identifying categories and themes of conversation within the data. Interpretations start with content analysis and thematic analysis (Braun and Clarke 2006, Elo and Kyngäs 2008) by coding contents in CADQAS software. The goal is to find relationships and common subjects within the clusters. This analysis considers the review of literature as a guide to link objectives with narratives and integrate sections, relationships and associations within the contents provided by the social networks (e.g., communities) (Boccaletti et al. 2006).

A content analysis develops the categories, where the themes are first searched in codes generated by the frequency of words. Likewise, with API, actors and actions (i.e. influencers), events and entities (i.e. organizations profiles) are identified, and conversation subjects and themes are used to construct narratives.

The information up to this point is not structured, but those common characteristics allow finding the categories. Qualitative analysis software – CADQAS used through groups of contents are organized first, by shared codes and words recurrent. This facilitates categorization by reading the shared content, thus it is possible to interpret it in thematic categories by content analysis (Braun and Clarke 2006), or a thick description when amounts of data provides several contents related (Lai and To 2015). This results in a corpus that once collected can be systematically analysed to understand and measure the information that people publish (publicly available), in social media and SNS to find common themes in conversations and interactions.

Each corpus of messages is a group of texts to interpret, in addition, according to the words, the positive, negative or neutral nature of the opinion can be discovered and thus even located in a compass related to the feelings within the contents (Dinakar et al. 2015). If the content is part of a dissemination or communication strategy usual on social networks, it is necessary to decide whether it is relevant. However, when each topic generates conversation, meaning the information sent is forwarded and receives responses, or there are replies between followers, originators or replicators, it is interesting because it is possible to find the meaning or nature of a topic. For example, in a theme about works of art, it is possible to find conversations about the experience in museums, the movements and organization of collectives, the media in which it is promoted, as well as the relationships between the labels used in the messages.

Next, we give examples of the behaviour of nodes by users, re-tweets and mentions, arising from the data of some studies. Firstly, we show the contents and interactions of users looking for activities in the cultural agenda of a city in Colombia. Secondly, we explain the result of discussions and content sharing about presidential elections. A third example shows all interactions about entrepreneurship that have been hash-tagged. The fourth and fifth examples show the interactions with comments about tourism providers and comments about experiences and mentions to an organization; the sixth shows a global community with the Pokémon Go App, which includes users talking about the places to find the creatures. Finally, the seventh example is about an event and its relationship with the city hosting it.

Examples of the Analysis of Complex Networks

Online communities are spaces for communication and interaction (Faraj et al. 2011), sometimes for collaboration (Rivera et al. 2010) or discussion (Valenzuela et al. 2012).

Kozinets (2002) explains that virtual spaces result from affiliations between individuals and groups sharing interests and knowledge, and engagement and involvement are part of the dynamics (Di Gangi and Wasko 2016) which allow the connectivity that explains a community online.

Communities also have engagement and narratives that influence and have effects on subsequent story-related attitudes and beliefs (Buselle and Bilandzic 2009). In that scenario, influencers manage engagement and spread information with their followers. Thus, Kozinets (2002) exposes that group gathering online has a purpose, sometimes leading to a high level of connections, dense interactions, and tight structures (Boccaletti et al. 2006).

To represent the process of applying the methodology, the treatment of data, Graph 1 shows the contents of Medellin's cultural agenda. The result shows the information flows that allow detecting information and interaction of the events in clusters of conversations (10 visible).

Each cluster acquires a colour in which contents are organized and interpreted.

For example, it detects that the Museum of Modern Art of Medellín makes appropriate use of its hashtag: #elMAMM to achieve a characterization of the marker. However, the node is dispersed if users tag the museum with markers such as #MAMM, #mamm, which in the searches lead to external topics or activities without any relation.

With this analysis, it is possible to find initiatives that generate their hashtag and encourage their audiences to share them to ensure that the content has an impact through collective action. This is the case of Proyecto NN a social

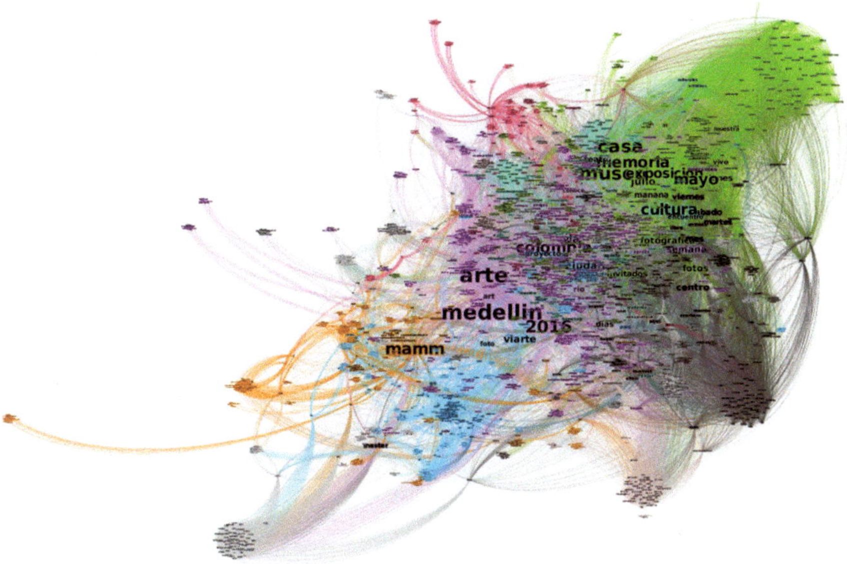

Graph 1: Cultural Communities in Medellín from their Thematic Axes
Source: API + Gephi (2017)

association that is using the hashtag #Aguanta (Hold!), generates actions by sharing, disseminating, and making the organization's purposes visible. The hashtag makes the initiative visible by linking different nodes (collectives or individuals) by inviting their followers to give their opinion and debate on the projects of the group; they also manage to develop their cartography by asking users to share geo-referenced content with the hashtag so that they can register actions and track opinions.

The rounded shape has variations that allow for the exploration of situations or problems. The circles represent the nodes and their relationships, as well as the curved lines which identify how words are connected, and the syntax and semantic relationships when similar or different meanings are recognised.

Graph 2 a rounded flower shape in which each petal is a discourse. This flower is a visualization of contents and discussions in the SNS, about Colombia's presidential elections.

It shows some nodes which were expected because they are informative, such as news, inscriptions and vote overseas. Some other nodes were about results from congress elections or coalitions. But one cluster with the word Venezuela stands out in terms of density, it is the biggest cluster. This cluster, once separated, showed several contents similar in structure and text inserted in each of the other clusters, in a repetitive way sharing negative messages, many times not related to the subject of conversations.

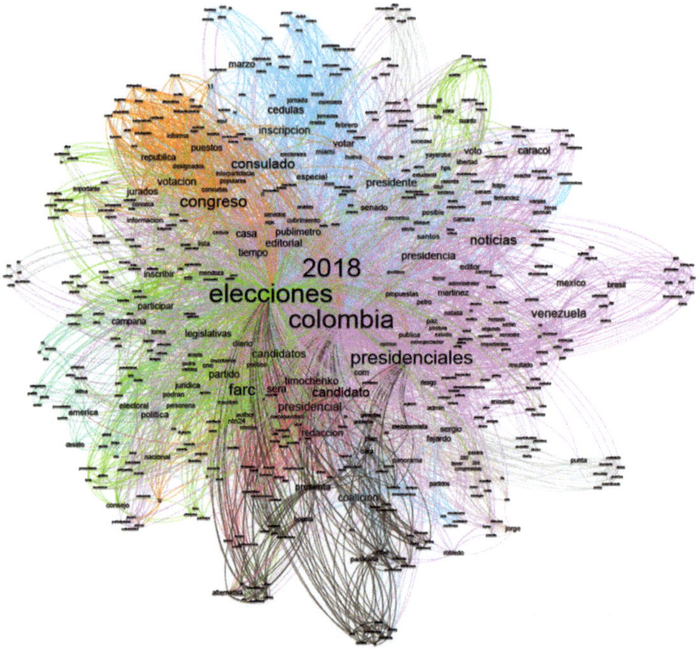

Graph 2: Presidential Elections
Source: API + Gephi

When the form is round-shaped it means that there is a highly connected community with a scale-free distribution because the elements composing it are diverse and the interactions are not local (Barabasi and Reka 1999). In this entrepreneurship network (Graph 3), the size and attachment indicates continuous movement and growth of the network. It happens when new nodes and connections represent interactions with several subjects of interest between users and entities (organizations). This behaviour demonstrates the cohesion and density around the main entrepreneurship node and denotes the centrality of the case. The complexity of this network threw 32 clusters and millions of contents shared in a microblog format. This search was just for about one week but the search in English extends the possibilities.

This example works to explain the network-intrinsic properties of Centrality, Cohesion and Density.

Then the network (Graph 4) compared shows more nodes. Cohesiveness is not evident, and centrality remains in an entity, which means that all conversations are coming from a source and the other relevant nodes (actors) have a link with it as followers. This case shows the system of tourism management in Medellin and the central actor is the promotion agency of the city and the region. This can be interpreted as they are doing their job emitting the messages and the receivers are interested and related to the subject, but feedback is not insured. This allows

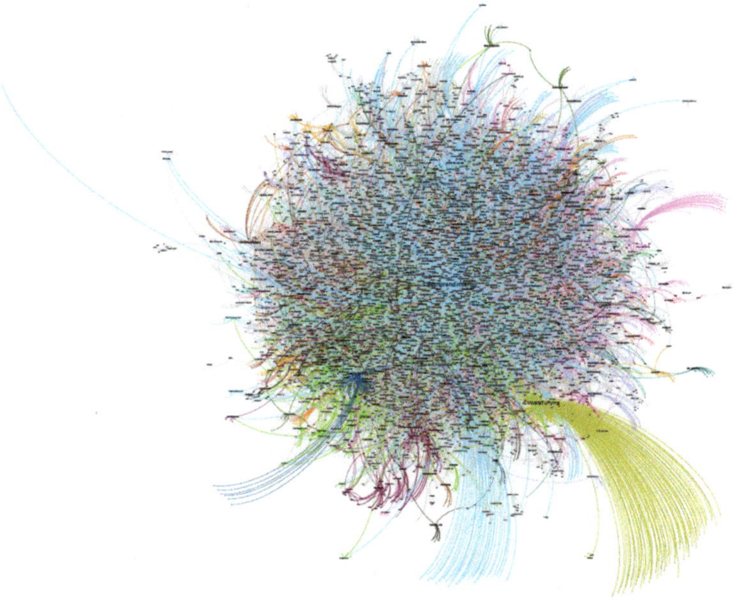

Graph 3: Entrepreneurship Conversation following the Hashtag
Source: API + Gephi

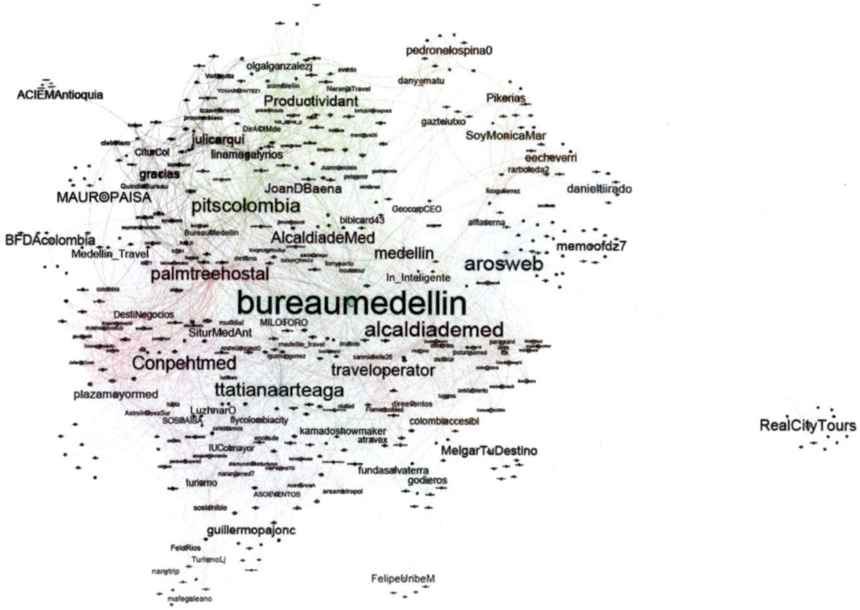

Graph 4: Medellin Tourism System
Source: API + Gephi

us also to explain the degree distribution (Estrada 2010), which is the tendency of nodes to group themselves around a central value and heterogeneity or level of irregularity. Then, nodes in each cluster have a degree of probability to be connected by an issue. But in this first exercise the search was made with a few words. Let's see what happens when the #Medellin is introduced and the search is extended as keyword, as in Graph 5.

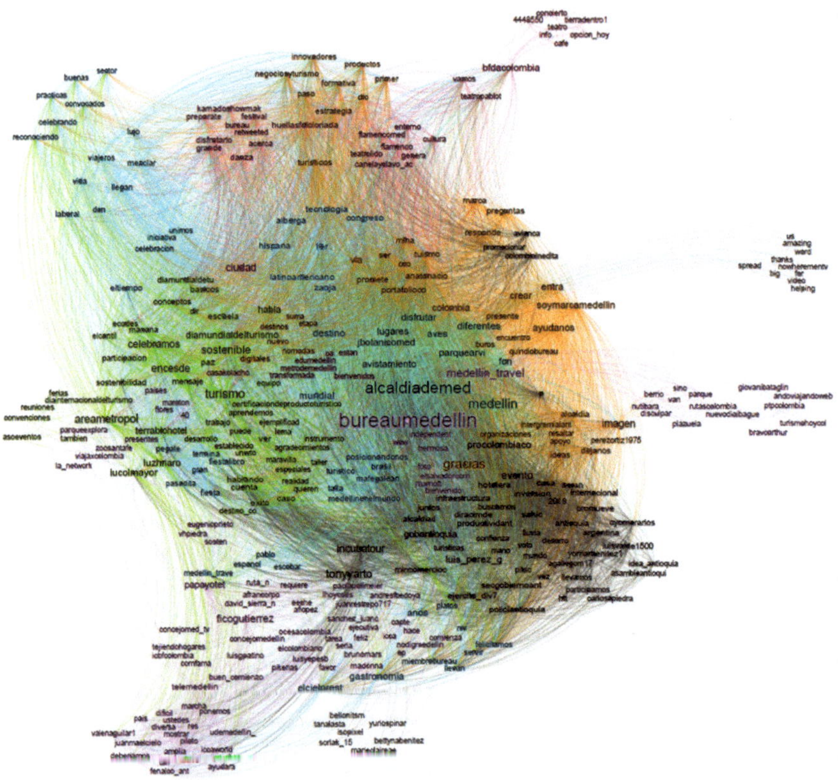

Graph 5: Medellin and the Greater Medellín Convention & Visitors Bureau in Tourism Users' Conversations
Source: API + Gephi

The topology of this network shows dense clusters when social media users are posting, commenting, and forwarding content within the community and the relationships in a social context. This visual analysis of the network according to the irregular shape explains the mechanics of interactions among users. In Graph 5 unstructured clusters with less cohesion but connectedness show conversations during an event, in this case the data is concentrated in periods near holidays.

This methodology aims to understand complex interaction patterns in SNA (Streeter and Gillespie 1993). Those interactions that connect actors (individuals, groups, and entities) sometimes concentrate on relational data. In the Pokémon Go (Graph 6) network analysis, the quality and quantity of associations and interdependencies show powerful social behaviour mapped as patterns, composed of points and lines in a mathematical space with formal properties (Crossley et al. 2009). Density does not allow it to do a network analysis. However, information on the internet is usually a complex network in which nodes are human beings and edges represent relations (Barabási 2002), which is why a complex network analysis and graphs are tools for depicting complexity. Thus, topology, geometric properties, and spatial relationships in graphs also demonstrate possible interactions. The tracking of nodes and vertices can be used to show communities composed of clusters. These classifications establish sub-graphs that are used to analyse the connections between network actors. Those 'layers' of the total network are useful to identify data, ordering items to describe, interpret and analyse subjects emerging from the interactions. Pokémon Go's analysis has 15 clusters, but is very dense as is shown in the next set of graphics.

Graph 6: Users' Conversations about Pokémon Go
Source: API + Gephi

Contents Extraction, Graph Analysis and Interpretation of Social Network Sites Users' Dynamic for Marketing Strategies

The use of analytics on data provided by users in SNS to study the content shared still requires deepening. This chapter presents the first bases used to

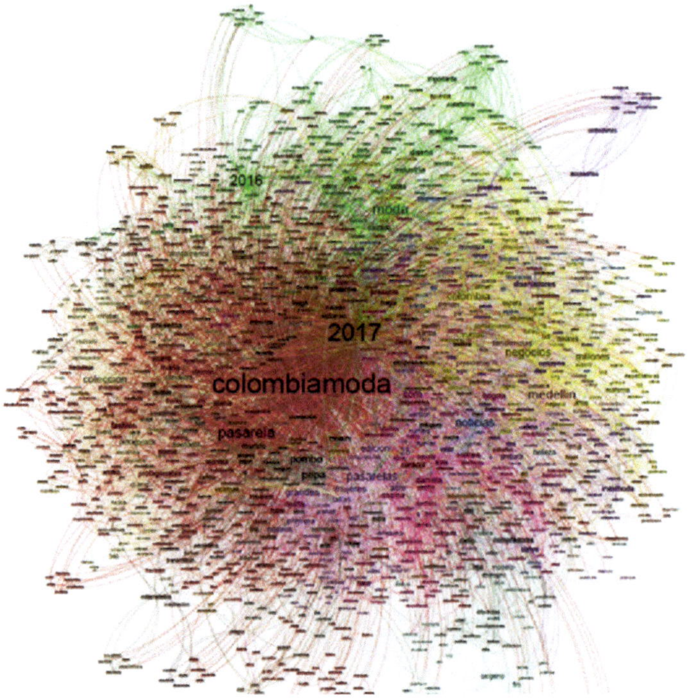

Graph 7: The Colombiamoda social network
Source: Ceballos et al. 2020

build a methodology to extract the topics that are discussed and shared on social networking sites in order to identify strategies that can guide marketing processes. For example, identifying not only influencers and their relationship with their followers but also the discourse they handle as shown in Graph 7.

This proposed methodology presents a data analysis process that utilizes various existing procedures with the aim to identify insights to reach consumers in digital environments, for example, with the graphical representation of a group of nodes (users) it is possible to know how cohesive or dispersed the community is. It is also possible to consider the users' needs by analysing their own conversations and online interactions when clustering in graphs, for example giving different colours to each cluster in order to find common codes and categories to interpret.

Data mining used to obtain a summary of the posts, messages and content shared on the SNS is an effective tool for studying information and analysing different contexts and topics. Also, it is possible to analyse the relationships among users with nodes and connections, and to identify the most influential profiles by analysing their participation in the conversation through the share of voice key performance indicator. By showing density in the graph of words (Heymann 2018), it is possible to follow and understand how users are commenting and interacting about topics of interest to find valuable contributions for proposing, revising or measuring communication, promotion, and marketing strategies.

In the methodological proposal we use a mix of different analyses and resources such as sentiment or opinion analysis, search algorithms, or machine learning. All of them, which contribute to the analysis of complex networks and to the dense description of a massive data set, contribute in different ways to finding and characterising user communities in a network system. Sometimes, when users share their interests with short texts (and sometimes accompanied by images or audio-visuals), the information published by them becomes a resource for study. A wide variety of elements to analyse are represented within them. Through data visualisation, for example, it is possible to detect key locations (or hotspots) when the information has geo-referenced data.

In short, posts that reflect the interaction of consumers in digital environments (commenting and discussing a topic), allow it to find codes and categories, explain topics, generate narratives, or show images that explain the meaning of content and discussions on social networks. At the same time, they illustrate a way of communicating with users using their own topics of conversation. The use of hashtags and keywords in the corpus of information analysed makes it possible to identify the data shared by users and classify it by topics. When formulating strategies, scrapped content itself cannot indicate relationships between discussion subjects, interactions among users, or the emergence or development of online communities and their complexity. Therefore, this type of online content is filtered and modelled, identifying, and classifying data using various analysis in the process.

This process let us to know how users are handling a topic in social networks; this also gives a basis for marketing strategies, measuring their effectiveness, understanding how this engages users and following the creation or expansion of communities online.

The chapter therefore summarises the procedure by which tweets are obtained, stored and analysed to determine and understand relationships and node formation from an identified and collected corpus of data. The analysis method allows us to solve several questions related to the predictive possibilities of networks. Firstly, it provides insight into the behaviour of users on a particular topic in social networks. Secondly, it helps to identify communities, nodes and relationships that develop an analysis of the impact of specific events. Finally, it facilitates the understanding of the communication and promotion strategies that are implemented on social networks, while making it possible to evaluate their effectiveness.

The visualisation methods outlined above aim to identify more effective ways of reaching user-consumers in digital environments used as channels, where not only is information provided, but also interaction with which to build a community and strengthen group relationships. Visual analysis of information also facilitates the identification of topics of conversation through hashtags and words. With this method it is also possible to examine and measure the communication strategy followed in social media as it makes evident the forms of direct participation of users in the conversation analysed.

The limitations of the methodologies presented are related to the collection processes and the forms of presentation of data shared by users. Much of this published data does not follow a prior structured procedure that can determine the type of data to be searched or analysed. Thus, the process of searching and working with the data, as they are not previously defined by variables, actions or profiles, is more complex and long-lasting.

Acknowledgements and Funding

This work was supported by an internal grant from Universidad EAFIT, Medellín, Colombia, for the project 'The cultural offer and the tourist experience in Medellín' (Code: 817245).

We appreciate the support provided by Sebastian Lopez (undergraduate auxiliary in research), as well as Lina Ceballos-Ochoa for proof reading in some contents.

References

Aparicio, S., J. Villazón-Terrazas and G. Álvarez. 2015. A model for scale-free networks: Application to twitter. *Entropy*, 17(8), 5848–5867. https://doi.org/10.3390/e17085848

Barabási, A.-L. 2002. *Linked: The New Science of Networks*. March 2003, 256. https://doi.org/10.2307/20033300

Barabási, A.-L. and R.E. Crandall. 2003. Linked: The New Science of Networks. *American Journal of Physics*, 71(4), 409–410. https://doi.org/10.1119/1.1538577

Boccaletti, S., V. Latora, Y. Moreno, M. Chavez and D.U. Hwang. 2006. Complex networks: Structure and dynamics. *Physics Reports*, Vol. 424(4–5), 175–308. North-Holland. https://doi.org/10.1016/j.physrep.2005.10.009

Braun, V. and V. Clarke. 2006. Using thematic analysis in psychology. *Qualitative Research in Psychology*, 3(2), 77–101. https://doi.org/10.1191/1478088706qp063oa

Brodie, R.J., A. Ilic, B. Juric and L. Hollebeek. 2013. Consumer engagement in a virtual brand community: An exploratory analysis. *Journal of Business Research*, 66(1), 105–114. https://doi.org/10.1016/j.jbusres.2011.07.029

Ceballos, L.M., L. Rojasdefrancisco and J.C. Monroy-Osorio. 2020. The role of a fashion spotlight event in a process of city image reconstruction. *Journal of Destination Marketing & Management*, 17(September 2019), 100464. https://doi.org/10.1016/j.jdmm.2020.100464

Chapman, P., J. Clinton, R. Kerber, T. Khabaza, T. Reinartz, C. Shearer and R. Wirth. 2000. Crisp-Dm 1.0. In: *CRISP-DM Consortium* (p. 76). https://doi.org/10.1109/ICETET.2008.239

Cinelli, M., G. de Francisci Morales, A. Galeazzi, W. Quattrociocchi and M. Starnini. 2021. The echo chamber effect on social media. *Proceedings of the National*

Academy of Sciences of the United States of America, 118(9). https://doi.org/10.1073/pnas.2023301118

Crossley, N., C. Prell and J. Scott. 2009. Social Network Analysis: Introduction to Special Edition. *Methodological Innovations Online*, 4(1), 1–5. https://doi.org/10.1177/205979910900400101

Davidson, T. and L. Farquhar. 2020. Prejudice and social media: Attitudes toward illegal immigrants, refugees, and transgender people. *Gender, Sexuality and Race in the Digital Age*, pp. 151–167. Springer International Publishing. https://doi.org/10.1007/978-3-030-29855-5_9

Di Gangi, P.M. and M.M. Wasko. 2016. Social media engagement theory: Exploring the influence of user engagement on social media usage. *Journal of Organizational and End User Computing*, 28(2), 53–73. DOI: 10.4018/JOEUC.2016040104

Dinakar, S., P. Andhale and M. Rege. 2015. Sentiment analysis of social network content. *Proceedings - 2015 IEEE 16th International Conference on Information Reuse and Integration, IRI 2015*, 189–192. https://doi.org/10.1109/IRI.2015.37

Elo, S. and H. Kyngäs. 2008. The qualitative content analysis process. *Journal of Advanced Nursing*, 62(1), 107–115. https://doi.org/10.1111/j.1365-2648.2007.04569.x

Erdõs, L., A. Knowles, H.T. Yau and J. Yin. 2013. Spectral statistics of Erdos-Rényi graphs I: Local semicircle law. *Annals of Probability*, 41(3 B), 2279–2375. https://doi.org/10.1214/11-AOP734

Estrada, E. 2010. Quantifying network heterogeneity. *Physical Review E – Statistical, Nonlinear, and Soft Matter Physics*, 82(6), 066102. https://doi.org/10.1103/PhysRevE.82.066102

Evans, D., S. Bratton and J. McKee. 2021. *Social Media Marketing*. AG Printing & Publishing. https://doi.org/10.5937/markt1704254k

Faraj, S., S.L. Jarvenpaa and A. Majchrzak. 2011. Knowledge collaboration in online communities. *Organization Science*, 22(5), 1224–1239. https://doi.org/10.1287/orsc.1100.0614

Förster, M., A. Mauleon and V.J. Vannetelbosch. 2016. Trust and manipulation in social networks. *Network Science*, Vol. 4(2), 216–243. Cambridge University Press. https://doi.org/10.1017/nws.2015.34

Frey, B.B. 2018. The SAGE Encyclopedia of Educational Research, Measurement, and Evaluation. *In*: *The SAGE Encyclopedia of Educational Research, Measurement, and Evaluation*. https://doi.org/10.4135/9781506326139

Furht, B. 2010. Handbook of Social Network Technologies and Applications. *In*: B. Furht (Ed.), *Springer*, Vol. 6(2). Springer.

Gephi. 2017. Gephi – The open graph viz platform. In Gephi. Gephi.org. https://gephi.org/ Retrieved 01.03.20 from https://gephi.org/

Glez-Peña, D., A. Lourenço, H. López-Fernández, M. Reboiro-Jato and F. Fdez-Riverola. 2013. Web scraping technologies in an API world. *Briefings in Bioinformatics*, 15(5), 788–797. https://doi.org/10.1093/bib/bbt026

Guliciuc, V. 2014. Complexity and Social Media. *Procedia – Social and Behavioral Sciences*, 149, 371–375. https://doi.org/10.1016/j.sbspro.2014.08.193

Hema, R. and N. Malik. 2010. Data Mining and Business Intelligence. *Proceedings of the 4th National Conference*.

Heymann, S. 2018. Encyclopedia of Social Network Analysis and Mining. *In*: R. Alhajj & J. Rokne (Eds.), *Encyclopedia of Social Network Analysis and Mining* (pp. 612–625). Springer. https://doi.org/10.1007/978-1-4939-7131-2

Kane, G.C., M. Alavi, G. Labianca and S.P. Borgatti. 2014. What's different about social media networks? A framework and research agenda. *MIS Quarterly: Management Information Systems*, 38(1), 275–304. https://doi.org/10.25300/misq/2014/38.1.13

Kaplan, A.M. and M. Haenlein. 2010. Users of the world, unite! The challenges and opportunities of Social Media. *Business Horizons*, 53(1), 59–68. https://doi.org/10.1016/j.bushor.2009.09.003

Kozinets, R.V. 2012. Marketing Netnography: Prom/ot(ulgat)ing a New Research Method. *Turkish Online Journal of Educational Technology*, 7(1), 37–45. https://doi.org/10.4256/mio.2012.004

Lai, L.S.L. and W.M. To. 2015. Content analysis of social media: A grounded theory approach. *Journal of Electronic Commerce Research*, 16(2), 138–152. http://www.jecr.org/node/466

Mangold, W.G. and D.J. Faulds. 2009. Social media: The new hybrid element of the promotion mix. *Business Horizons*, 52(4), 357–365. https://doi.org/10.1016/j.bushor.2009.03.002

McGloin, J.M. and D.S. Kirk. 2010. Social network analysis. *In*: A. Piquero & D. Weisburd (Eds.), *Handbook of Quantitative Criminology* (pp. 209–224). Springer.

Medellin Bureau. 2021. Bureau - 2020. Bureaumedellin. https://www.bureaumedellin.com/

Miettinen, S. 2010. Service design: New methods for innovating digital user experiences for leisure. *In*: *Digital Culture and E-Tourism: Technologies, Applications and Management Approaches* (pp. 36–47). IGI Global. https://doi.org/10.4018/978-1-61520-867-8.ch003

Moqbel, M. and N. Kock. 2018. Unveiling the dark side of social networking sites: Personal and work-related consequences of social networking site addiction. *Information and Management*, 55(1), 109–119. https://doi.org/10.1016/j.im.2017.05.001

Rivera, M.T., S.B. Soderstrom and B. Uzzi. 2010. Dynamics of dyads in social networks: Assortative, relational, and proximity mechanisms. *Annual Review of Sociology*, 36, 91–115. https://doi.org/10.1146/annurev.soc.34.040507.134743

Roma, P. and D. Aloini. 2019. How does brand-related user-generated content differ across social media? Evidence reloaded. *Journal of Business Research*, 96, 322–339. https://doi.org/10.1016/j.jbusres.2018.11.055

Saurkar, A.V., K.G. Pathare and S.A. Gode. 2018. An Overview on Web Scraping Techniques and Tools. *International Journal on Future Revolution in Computer Science & Communication Engineering*, 4(4), 363–367. http://www.ijfrcsce.org

Scott, J. 2011. Social network analysis: Developments, advances, and prospects. *Social Network Analysis and Mining*, 1(1), 21–26. https://doi.org/10.1007/s13278-010-0012-6

Smith, A.N., E. Fischer and C. Yongjian. 2012. How does brand-related user-generated content differ across YouTube, Facebook, and Twitter? *Journal of Interactive Marketing*, 26(2), 102–113. https://doi.org/10.1016/j.intmar.2012.01.002

Sonawane, S. and P.A. Kulkarni. 2014. Graph based representation and analysis of text document: A survey of techniques. *International Journal of Computer Applications*, 96(19), 1–8. https://doi.org/10.5120/16899-6972

Stephen, A.T. 2016. The role of digital and social media marketing in consumer behavior. *Current Opinion in Psychology*, 10, 17–21. https://doi.org/10.1016/j.copsyc.2015.10.016

Streeter, C.L. and D.F. Gillespie. 1993. Social network analysis. *Journal of Social Service Research*, 16(1–2), 201–222. https://doi.org/10.1300/J079v16n01_10

Thong, J.Y.L., V. Venkatesh, X. Xu, S.J. Hong and K.Y. Tam. 2011. Consumer acceptance of personal information and communication technology services. *IEEE Transactions on Engineering Management*, 58(4), 613–625. https://doi.org/10.1109/TEM.2010.2058851

Valenzuela, S., Y. Kim and H. Gil De Zúñiga. 2012. Social networks that matter: Exploring the role of political discussion for online political participation. *International Journal of Public Opinion Research*, 24(2), 163–184. https://doi.org/10.1093/ijpor/edr037

Watts, D. 2003. Six degrees: The science of a connected age. *Journal of Marketing*, 68(1), 166–167. https://doi.org/loc?

Watts, D.J. and S.H. Strogatz. 1998. Collective dynamics of small-world9 networks. *Nature*, 393(6684), 440–442. https://doi.org/10.1038/30918

Part III

Ethical and Political Implications of Prediction

CHAPTER
6

Predicting Government Attention in Social Media: A First Step for Understanding Political *astroturf* in Interest Representation

Camilo Cristancho Mantilla[1]
University of Barcelona, Avinguda Diagonal, 684, 08034 Barcelona (Spain)

Introduction

Government attention towards political issues is a crucial matter both in terms of policy making and democratic responsiveness. The needs and wishes of citizens are transmitted through social actors who suggest which issues will be considered and how social demands will be defined. Which actors get heard and whose definitions of social issues are publicly accepted is a central concern for democratic principles. Understanding this preference transmission process is especially important in a rapidly changing digital arena. Considering that the salience and interactions between actors in social media which are governed to a large extent by the corporate interests of social media platforms and are susceptible to the manipulation of public opinion, it is crucial to recognize how the government picks up on this social signal. This chapter raises questions on to what extent government attention can be predicted based on the social media activity of formal social actors. The extent in which the issue definitions used by social actors on social media predict government attention signals a healthy representative system which remains shielded from devices aimed at manipulating

[1] This publication is part of the I+D+i project RTI2018-100861-J-I00, funded by MCIN/AEI 10.13039/501100011033 and "FEDER Una manera de hacer Europa".

Email: camilo.cristancho@ub.edu

public opinion. This is particularly relevant considering the increasing claims of an unresponsive democracy and the increased threats of artificial manipulation of public opinion.

This chapter describes the extent to which the government and presidential agenda can be predicted by the issue of attention of interest groups in Spain using Twitter data from the 150 most active interest group organisations and the official users from the government cabinet between March 2018 and March 2021.

An aggregate analysis for all types of interest group organisations shows that the presidential agenda can be better predicted than the government agenda by the activity of interest group organisations. When looking into the different types of interest group organisations, the analysis confirms already established findings on the differential attention of government to interest groups. Business and special interest group's organisations, as well as unions are the only types of interest groups that predict the presidential agenda. The effect of unions is quite large when considering temporal dependencies, but unions are also the only type of organisation that loses its predictive power over time. These findings have implications for the problem of regulating the use of AI (artificial intelligence) by interest groups. The fact that linkages between government and presidential agendas with interest group agendas can be mapped is a first step which provides some optimism on the ability to control the use of disruptive technologies like AI.

AI and Political Processes

The effects of artificial intelligence (AI from here on) on democracy have raised all kinds of expectations and fears. Political processes no longer depend only on human behaviour but are partly dependent on systems with extraordinary capabilities for data processing and results which are not fully predictable. AI involves multiple technologies that appear to act as if they were rational beings (Turing 1950). AI technologies involve the use of algorithms, machine learning, feedback systems, and automation in order to simply perform the functions that they were programmed for. AI has the potential to disrupt how we conceive democratic self-government and how we make political decisions. The automation of political communication raises various questions directly related to the normative core of democracy. How are advanced communication techniques and data analytics affecting political mobilisation? Who expresses public preferences in automated environments? To what extent is public opinion being artificially managed?

The transformation of digital communications and public arenas are forcing us to rethink some of the basic categories of politics and the existing decision processes. Sophisticated and complex technologies that power learning machines, data analytics in gigantic proportions or the current proliferation of automated decision systems are not devices that can be regulated with simple intervention procedures. A first step is to characterise and determine the functioning of the regular political processes in order to assess the magnitude of the potential social threats brought about by the use of AI in these processes.

Concerns about artificial intelligence have dealt with its ethical, legal, and economic dimensions, but have focused less on its political dimension. Although there is growing awareness regarding manipulated content on social media, survey data shows that a large majority of European citizens are supportive of voting in elections through their smartphone (72%) and more than half support giving seats of national parliamentarians to an algorithm (51%) (Jonsson and Luca de Tena 2021). The role of social organisations in politics is largely ignored when considering social perceptions of digital campaigns. However, campaigns based on ultra-narrowcasting technologies which are increasingly successful in targeting customized political messages with an unprecedented precision for communicating specific contents with tailored emotional charges are quite present.

The Brexit referendum and the 2016 Presidential Election in the United States are the most notorious cases in which the integrity of the campaigns has been questioned. The involvement of data analytics firms such as Cambridge Analytica, which allegedly employed advanced AI technologies to manipulate the electorate is a major public concern. Related claims have resulted in investigations in the UK into the possible illegal use of data by the Information Commissioner's Office and unlawful donations to the official Leave campaign by the Electoral Commission. However, these have not been the only cases of shady interference in political processes using AI.

In 2016, Colombia held a referendum to ratify the peace agreement with the Revolutionary Armed Forces of Colombia (FARC). The polls largely predicted the victory of the peace agreement with a comfortable margin. Surprisingly, the peace agreement was defeated (Gómez-Suárez 2016). Days later, the organisers of the campaign proudly explained how they used narrowcasting techniques to instil anger and fear to mobilize opposition to the peace agreement[2].

The atypicality of these cases raises questions on how to explain these unexpected results. Although research has explained the complexity of voting referendums based on predispositions and ideological basis rather than on the actual question (DeVreese and Semetko 2004), several claims on the fraudulent manipulation of the campaigns provide an alternative explanation.

Research on algorithmic decision-making has focused on how the content on social media is produced by computer programs that use massive quantities of detailed data to deliver content that users find relevant or engaging. Public concerns for misinformation, polarisation, or hate-speech blame the platforms for promoting harmful content. However, these types of campaigns use the platforms' functionalities, but are orchestrated and financed by organised interests.

Social media platforms have also played a role in facing these challenges. In October 2018, in the context of the so-called Catalan Referendum, Twitter

[2] The interview was published in the national daily *El Colombiano* on October 6th, 2016. http://www.elcolombiano.com/colombia/acuerdos-de-gobierno-y-farc/entrevista-a-juan-carlos-velez-sobre-la-estrategia-de-la-campana-del-no-en-el-plebiscito-CE5116400

suspended 130 fake accounts associated with the Catalan party Esquerra Republicana de Catalunya. These accounts were used to influence the conversation in politically advantageous ways and spread content about the Catalan independence movement. Six months later, 259 accounts associated with the main opposition party, Partido Popular, were also suspended for spamming or retweeting content to increase engagement. They were suspected of falsely boosting public sentiment online in the context of the general elections in Spain. This is part of an initiative led by Twitter covering coordinated, state-backed activities. They argue that those with the advantages of institutional power who consciously abuse the rules of the platform are "not advancing healthy discourse but are actively working to undermine it" (Twitter Safety 2019). This type of initiative aimed at identifying the malicious use of AI is one of the few ways to further our understanding on issues which threaten the integrity of public conversation online and undermine the logics of public arenas and interest representation.

Academic research on the political implications of AI has looked into computational propaganda – the use of social media platforms, actors, and big data for the manipulation of public opinion (Woolley and Howard 2018), and astroturf, the inappropriate use of technology for forging political mobilisation and the deceitful influence on public opinion. In reference to a surface of synthetic fibres which intends to look like natural grass, astroturf campaigns are not grassroots, but manipulated to back claims about an issue in a way that is deliberatly disguising its origins (Lock et al. 2016).

Multiple investigations on the external involvement and disruption in national elections have looked into the influence of bots on the US Presidential elections (Kollanyi et al. 2016), Venezuelan elections (Forelle et al. 2015), or Russian bots being involved in the politics of Western countries (Barash and Kelly 2012), and the Chinese governmental controls on social media (King et al. 2013, 2017). Research on the social perspective of computational propaganda has focused on the use of bots and the use of big data in politics (Woolley and Howard 2018).

However, political mobilisation is not limited to referendums or electoral campaigning. In the context of interest group representation, astroturf lobbying has been defined as a deceiving lobbying practice that undermines and fakes grassroots movements. The impact of lobbying activity has a larger disruptive potential as it is responsible for policy definition and the mobilisation of public support as a continuous process in the legislative and regulatory action. Furthermore, its routine actions can be concealed easily as they are not in the radar of public attention in the scale of elections or public referendums.

This chapter argues that detecting astroturf lobbying implies recognizing unexpected behaviours in the patterns of interest representation in a similar manner than the unexpected results in electoral or referendum campaigns that have triggered the alarms of malicious practices. In order to do so, it discusses the literature on interest representation and astroturf lobbying and presents a baseline for predicting government attention using interest groups agendas on Twitter.

Government Attention and Interest Groups

The extent to which policymakers focus their attention on issues that are important to the public is a central premise of democratic representation (Bevan and Jennings 2014). Research shows that policymakers tend to respond to issues that are important to the public (Burstein and Linton 2002). However, even if elected officials face strong incentives to listen to voters, these interests are mediated by multiple types of organisations that use multiple opportunities and spaces for extra parliamentary representation. Parties are closely connected to multiple social movement organisations, they listen to social demands on the street, and they seek the information provided by lobbyists as these provide critical resources (Downs 1972). Interest groups, loosely defined as organisations that attempt to influence public policy, play a central role in agenda setting (Schattschneider 1960). They work to highlight or define policy problems, mobilize public support, push issues or hide them from public attention, and provide information to legislators (Baumgartner et al. 2008, 2014, Baumgartner and Leech 1998). This makes interest groups central players in shaping policy agendas. IG organisations bring citizens' concerns to the government and enabling their influence on the policy agenda (Dahl 2005). They mobilize interests by prompting representatives to interact and become increasingly aware of these interests, and they also challenge status quo arrangements by reframing policy and breaking policy monopolies (Jones et al. 1993).

However, not all interests are equal in their influence capacity, and this leads to differential access and power (Binderkrantz et al. 2014, Lindblom 1977). This implies high risks of corruption and public perceptions of undue influence. As a result, governments have a long tradition in developing transparency mechanisms such as regulating lobbying procedures and holding interest groups accountable for their activities (Holman and Luneburg 2012). Notwithstanding, these efforts run short when facing modern communication practices which involve campaigning and recruiting methods powered by AI techniques. The use of AI not only involves the potential misuse of technology for particular interests that go against the common interest, but also by the potential inequalities brought about by the fact that the development of AI is dominated by the most powerful organisations. The largest organisations may afford larger investment in the area of AI, and they can pay as well for privileged access to the infrastructures of public discourse and the digital environment decisive for political debate where public campaigns, policy definition and elections take place. This implies as well the outsourcing communication processes, but also efficiently collecting personal data for profiling their members, followers and those that oppose them.

Considering this scenario, understanding the ability of organisations for shaping government attention to particular issues is a first step for measuring their influence and setting the required baseline for identifying atypical or suspicious behaviour which signals the use of technological advantages which may be infringing the ethical conventions of interest group representation.

Empirical Approach

The growing use of social media by political actors has brought hope into the potential transformation of politics in an unrestricted arena with no gatekeepers, and with low levels of institutional restrictions and hierarchical control such as the parliamentary and media venues. However, research has found that parties and politicians use online communication mostly for spreading news about themselves (Golbeck et al. 2010), replicating messages from other platforms (Larsson 2015), and expressing issue positions that amplify their party's message (Kruikemeier 2014). Politicians and parties emphasize different issues to different audiences across platforms in order to better cater their message to the public (Jungherr et al. 2016, Kreiss 2014). Research on interest groups, social movement organisations and Unions have also pointed to their increased uptake of social media in all their organisational activity (Kanol and Nat 2017).

In order to measure the issue of attention of government and interest groups on social media, the empirical analysis is based on tweets from 22 official Twitter users of the government cabinet and the president, and a sample of 149 interest group organisations. These were identified using two sources: a list of the groups with the longest experience in their interactions with government (Chaqués-Bonafont et al. 2021) and the groups that have appeared in the reconstruction commission of the Spanish parliament (as a response to the Covid pandemic) in 2020. A total of 166 groups have been identified using these two lists. Seventy of these are on both lists, and the final sample consists of 149 organisations which have a Twitter handle. The main handle of each organisation is used, except for an organisation that has no national representation but 10 sub-national handles (*Las Kellys*). It is important to note that the sample has a marked bias towards the issues most affected by the pandemic on behalf of the organisations selected from their appearance in the reconstruction commission. This mainly includes business associations and unions of public employees in health and education, small and medium-sized enterprises (SMEs), and student organisations and farmers unions. The implications of this bias are discussed in the presentation of the results.

For these 149 users, 536,420 tweets have been collected between March 1, 2018, and March 1, 2021 (Figure A1). The interest group sample is made up of five types of groups that are distributed in special interest (38%), NGOs (23%), Business (13.8%), Professional associations (8.2%) and Unions (17%). The sample for official government and cabinet users is composed of 75,956 tweets in the same timeframe.

Interest Representation, Disinformation and Astroturf

Political astroturfing has been defined as "hidden propaganda by powerful political actors aimed at mimicking grassroots activity, on social media" (Keller et al. 2017). This implies that astroturfing is hard to distinguish from genuine grassroots support as it involves complex techniques for modifying content and communication patterns in a way that seems natural. The mechanisms for detecting astroturfing mostly rely on identifying suspicious accounts and content or unusual communication patterns such as abnormal activity thresholds (Howard and Kollanyi 2016). This means that studies need cases in which they can correctly learn how "normal patterns" look and cases in which atypical patterns have been identified.

The purpose of this article is to measure the extent in which the government agenda can be predicted by the interest groups' agenda. The government agenda is composed of the issues that receive attention by the cabinet leaders or the press offices of ministries at a given point in time. This is a proxy for issues that are considered important or in need of attention (Cobb and Elder 1972, Kingdon 1993). Similarly, the interest group agenda is composed by the aggregate attention of organisations in each of five interest group types: Business organisations, NGOs, Professional associations, Unions, and Special interest organisations.

The first step is characterizing the agendas by measuring how interest groups and ministries distribute their attention between policy issues on Twitter. Tweets by 149 interest group organisations, 22 ministries and a handle which represents the presidential agenda are classified into 20 policy issues following the issue categories of the Comparative Agendas Project (CAP) (Baumgartner et al. 2019).

A supervised automatic classification process is used to assign each tweet to the corresponding policy category. The classifier is trained (learns the classification criteria) using the text of press releases published by the ministries during the period of analysis (03-01-2018 to 03-01-2021, N = 4,089) in such a way that the unique terms that compose a representative pattern of each of the 20 CAP categories associated with each ministry are extracted[3]. This approach allows thousands of tweets to be classified fully and automatically in just a few minutes. Classifier performance statistics show that the quality of the automatic classification process is excellent with 96% accuracy and the variation between categories is small.

The distribution of issue attention by actor type is presented in Figure 6.1.

In order to predict the government's issue attention, all data are aggregated in a pooled time-series, with weeks nested in issue categories. This means that the observation is the weekly share of attention for each of the issue categories for each agenda. Both the independent (interest groups' agenda) and the dependent (government and presidential agendas) variables refer to the proportion of attention to a given issue on a weekly basis.

Agenda-setting studies assert that the relationship between government attention to policy issues and the attention of other actors is reciprocal.

[3] Press releases published in the government website: https://www.lamoncloa.gob.es/serviciosdeprensa/notasprensa/

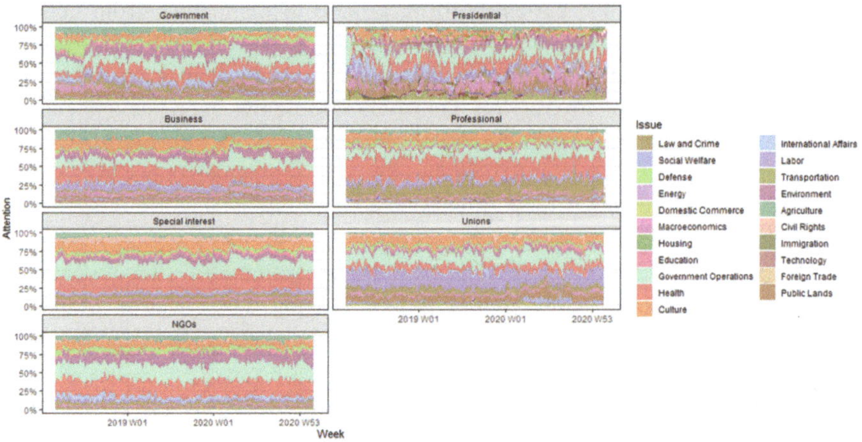

Figure 6.1: Issue Attention

Government actors adopt issues that are prioritized by social organisations, but the representation process also has feedback mechanisms in which social organisations react to government initiatives and to issues that have already received political attention.

The government and presidential agendas are predicted using a pooled time-series structure. This has three implications: In the first place, to describe a causal effect in which the government/president follows interest groups, the cause has to precede the consequence. This is modelled by using lagged values for the independent variable in the prediction. In the second place, the momentum of the government and presidential agenda is expected to have a strong predictive power. A lagged dependent variable is used to deal with this temporal dependency (autocorrelation). In the third place, there is a need to control the heterogeneity of inter-issue variation. Dummy variables for each issue are introduced in the model (fixed effects).

Results

The predictability of the government and presidential attention is tested by the aggregate agenda of interest groups (Table 6.1) and the agenda of each type of group (Table 6.2). Both tests include models with the main effects of the interest groups agenda, and an additional model that includes the time interaction. This interaction shows how the effect of the interest groups' agenda on the presidential and government agendas changes in time.

Results show the significant positive effects of the interest group agenda on the presidential agenda at an aggregate level (Models 1 and 3 in Table 6.1), and for the special interest groups and the business organisations in the presidential agenda (Model 1 in Table 6.2). The effects are small when compared to autoregressive elements that confirm the inertia of the agendas. However, the effect of the

Table 6.1: Predicting the Government and Presidential Agendas – Aggregate Model

	Presidential agenda	Government agenda	Presidential agenda + Time FE	Government agenda + Time FE
Interest group agenda (lag)	0.195*** (0.072)	0.020 (0.035)	4.309*** (1.068)	-0.004 (0.484)
Presidential agenda (lag)	0.306*** (0.023)		0.297*** (0.023)	
Government agenda (lag)		0.572*** (0.017)		0.571*** (0.017)
YearWeek			0.00001** (0.00001)	-0.00000 (0.00000)
Interest groups' agenda (lag)*YearWeek			–0.0002*** (0.0001)	0.00000 (0.00003)
Constant	0.046*** (0.005)	0.019*** (0.002)	–0.157 (0.102)	0.031 (0.040)
Observations	1,853	2,496	1,853	2,496
R^2	0.412	0.761	0.418	0.761

Note: *p<0.1; **p<0.05; ***p<0.01
Fixed effects for issues not shown.

Union's agenda on the presidential agenda is quite strong (Model 3 in Table 6.2). The interest groups agenda loses its predictive capacity on the presidential agenda with time (temporal fixed effect in Model 3 in Table 6.1), but this effect seems to be exclusive for Unions (temporal fixed effect in Model 3 in Table 6.2).

Conclusion

In famous cases of public opinion manipulation in the last ten years have been widely discussed in context of their critical implications. Regulators thus face the need to preventively limit or even ban certain applications out of concern for the worst-case scenarios. Conversely, they do not want to obstruct the development of a technology that underlies a rapidly growing part of the economy. The decision should depend on the ability to determine with great precision the ways and scenarios in which fundamental democratic rights are challenged by the use of AI and need to be protected accordingly. This chapter contends that predicting government attention using the social media activity of formal social actors is useful as a baseline for monitoring the potential threats of the malicious use of AI technologies in the policy process. Having a baseline to predict government attention provides a useful description of the expected behaviour of political

Table 6.2: Predicting the Government and Presidential Agendas by Interest Group Type

	Presidential agenda	**Government agenda**	**Presidential agenda + Time FE**	**Government agenda + Time FE**
Business orgs' agenda (lag)	0.097*	0.031	-0.508	0.109
	(0.052)	(0.025)	(1.877)	(0.880)
NGOs' agenda (lag)	0.025	–0.034	–1.602	–1.433
	(0.057)	(0.027)	(1.968)	(0.945)
Prof. associations' agenda (lag)	0.040	0.040**	–0.207	–0.144
	(0.039)	(0.018)	(1.283)	(0.587)
Unions' agenda (lag)	–0.013	–0.007	3.863***	0.355
	(0.045)	(0.021)	(1.145)	(0.552)
Special interest orgs' agenda (lag)	0.177**	0.039	4.108	1.086
	(0.069)	(0.034)	(2.549)	(1.187)
Presidential agenda (lag)	0.316***		0.300***	
	(0.023)		(0.023)	
Government agenda (lag)		0.576***		0.575***
		(0.018)		(0.018)
YearWeek			0.0001**	–0.00001
			(0.00005)	(0.00002)
Business orgs' agenda (lag)*YearWeek			0.0002	–0.00003
			(0.001)	(0.0003)
NGOs' agenda (lag)*YearWeek			0.001	0.001
			(0.001)	(0.0004)
Professional associations' agenda (lag)*YearWeek			0.0001	0.0001
			(0.0005)	(0.0002)
Unions' agenda (lag)*YearWeek			–0.001***	–0.0001
			(0.0004)	(0.0002)
Special interest orgs' agenda (lag)*YearWeek			–0.002	–0.0004
			(0.001)	(0.0005)
Constant	0.040***	0.017***	–0.239**	0.038
	(0.005)	(0.002)	(0.119)	(0.049)
Observations	1,795	2,306	1,795	2,306
R^2	0.408	0.743	0.417	0.743

Note: *p<0.1; **p<0.05; ***p<0.01
Fixed effects for issues not shown.

actors in the policy process and consequently the possibility of observing atypical behaviour, such as astroturf campaigns. Obviously, a naïve prediction model with a small sample of a single country's actors is only a starting point. However, it works as a proof-of-concept for using easily available data to describe the typical behaviour of the interest representation process in a normatively important issue such as agenda-setting.

Further research is needed in order to fully understand government attention and how the context matters. The interest group agenda is only one of multiple inputs into the attention of political representatives, and multiple additional venues and actors such as the media, the legislative process and public opinion should be considered for an improved model of government attention. More importantly, recurring data on the widespread adoption of AI technologies is extensively produced through business surveys and expense records. This data should also shed light on how AI is normalized in the policy process. A growing use of AI should eventually lead to the regulation and control of political organisations in a similar manner as they are for their campaign finances and their communication practices during special decision moments such as political campaigns or referendums, but also in the daily processes of issue representation.

References

Barash, V. and J. Kelly. 2012. Salience vs. commitment: Dynamics of political hashtags in Russian Twitter. Berkman Center Research Publication, 2012–9.

Baumgartner, F.R. and B.L. Leech.1998. *Basic Interests*. Princeton University Press.

Baumgartner, F.R., S.L. De Boef and A.E. Boydstun. 2008. *The Decline of the Death Penalty and the Discovery of Innocence*. Cambridge University Press.

Baumgartner, F.R., J.M. Berry, M. Hojnacki, D.C. Kimball and B.L. Leech. 2014. Money, priorities, and stalemate: How lobbying affects public policy. *Election Law Journal*, 13(1), 194–209.

Baumgartner, F.R., C. Breunig and E. Grossman. 2019. *Comparative Policy Agendas: Theory, Tools, Data*. Oxford University Press.

Bevan, S. and W. Jennings. 2014. Representation, agendas and institutions. *European Journal of Political Research*, 53(1), 37–56. https://doi.org/10.1111/1475-6765.12023

Binderkrantz, A.S., P.M. Christiansen and H.H. Pedersen. 2014. A privileged position? The influence of business interests in government consultations. *Journal of Public Administration Research and Theory*, 24(4), 879–896.

Burstein, P. and A. Linton. 2002. The impact of political parties interest groups and social movement organizations on public policy. *Social Forces*, 81(2), 381–408.

Chaqués-Bonafont, L., C. Cristancho, L. Muñoz and L. Rincón. 2021. The Contingent Character of Interest Groups-Political Parties' Interaction. *Journal of Public Policy*, 41(3), 440–461.

Cobb, R.W. and C.D. Elder. 1972. The dynamics of agenda-building. *Classics of Public Policy*, 128–136.

Dahl, R.A. 2005. *Who Governs?: Democracy and Power in an American City*. Yale University Press.

DeVreese, C.H. and H.A. Semetko. 2004. News matters: Influences on the vote in the Danish 2000 euro referendum campaign. *European Journal of Political Research*, 43(5), 699–722.

Downs. 1972. The issue – Attention cycle. *The Public Interest*, 28, 38–50.

Forelle, M., P.N. Howard, A. Monroy-Hernández and S. Savage. 2015. Political bots and the manipulation of public opinion in Venezuela. Available at SSRN 2635800.

Golbeck, J., J.M. Grimes and A. Rogers. 2010. Twitter use by the US Congress. *Journal of the American Society for Information Science and Technology*, 61(8), 1612–1621.

Gómez-Suárez, A. 2016. *El triunfo del No: la paradoja emocional detrás del plebiscito*. Ícono.

Holman, C. and W. Luneburg. 2012. Lobbying and transparency: A comparative analysis of regulatory reform. *Interest Groups & Advocacy*, 1(1), 75–104.

Howard, P.N. and B. Kollanyi. 2016. Bots,# stronger in, and# brexit: Computational propaganda during the UK-EU referendum. Available at SSRN 2798311.

Jones, B.D., F.R. Baumgartner and J.C. Talbert. 1993. The destruction of issue monopolies in Congress. *American Political Science Review*, 87(3), 657–671.

Jonsson, O. and C. Luca de Tena. 2021. European Tech Insights: Mapping European Attitudes towards Technological Change and its Governance. https://www.ie.edu/cgc/research/european-tech-insights/

Jungherr, A., H. Schoen and P. Jürgens. 2016. The mediation of politics through Twitter: An analysis of messages posted during the campaign for the German Federal Election 2013. *Journal of Computer-Mediated Communication*, 21(1), 50–68. https://doi.org/10.1111/jcc4.12143

Kanol, D. and M. Nat. 2017. Interest groups and social media: An examination of cause and sectional groups' social media strategies in the EU. *Journal of Public Affairs*, September 2016, e1649. https://doi.org/10.1002/pa.1649

Keller, F.B., D. Schoch, S. Stier and J. Yang. 2017. How to manipulate social media: Analyzing political astroturfing using ground truth data from South Korea. *Proceedings of the 11th International Conference on Web and Social Media, ICWSM 2017*, *Icwsm*, 564–567.

King, G., J. Pan and M.E. Roberts. 2013. How censorship in China allows government criticism but silences collective expression. *American Political Science Review*, 107(2), 326–343.

King, G., J. Pan and M.E. Roberts. 2017. How the Chinese government fabricates social media posts for strategic distraction, not engaged argument. *American Political Science Review*, 111(3), 484–501.

Kingdon, J.W. 1993. How do issues get on public policy agendas. *Sociology and the Public Agenda*, 8(1), 40–53.

Kollanyi, B., P.N. Howard and S.C. Woolley. 2016. Bots and automation over Twitter during the first US presidential debate. *Comprop Data Memo*, 1, 1–4.

Kreiss, D. 2014. Seizing the moment: The presidential campaigns' use of Twitter during the 2012 electoral cycle. *New Media & Society*. https://doi.org/10.1177/1461444814562445

Kruikemeier, S. 2014. How political candidates use Twitter and the impact on votes. *Computers in Human Behavior*, 34, 131–139.

Larsson, A.O. 2015. The EU Parliament on Twitter—Assessing the permanent online practices of parliamentarians. *Journal of Information Technology & Politics*, 12(2), 149–166.

Lindblom, C.E. 1977. Politics and markets: The world's political. *Economic Systems*, 9.

Lock, I., P. Seele and R.L. Heath. 2016. Where grass has no roots: The concept of 'shared strategic communication' as an answer to unethical astroturf lobbying. *International Journal of Strategic Communication*, 10(2), 87–100.

Schattschneider, E.E. 1960. *Party Government.* Transaction Publishers.

Turing, A.M. 1950. Computing Machinery and Intelligence. *MIND*, *LIX* (236), 433–460.

Twitter Safety. 2019. Disclosing new data to our archive of information operations. Twitter. https://blog.twitter.com/en_us/topics/company/2019/info-ops-disclosure-data-september-2019 Blog. https://blog.twitter.com/en_us/topics/company/2019/info-ops-disclosure-data-september-2019

Woolley, S.C. and P.N. Howard. 2018. *Computational propaganda: Political Parties, Politicians, And Political Manipulation On Social Media*. Oxford University Press.

Appendix

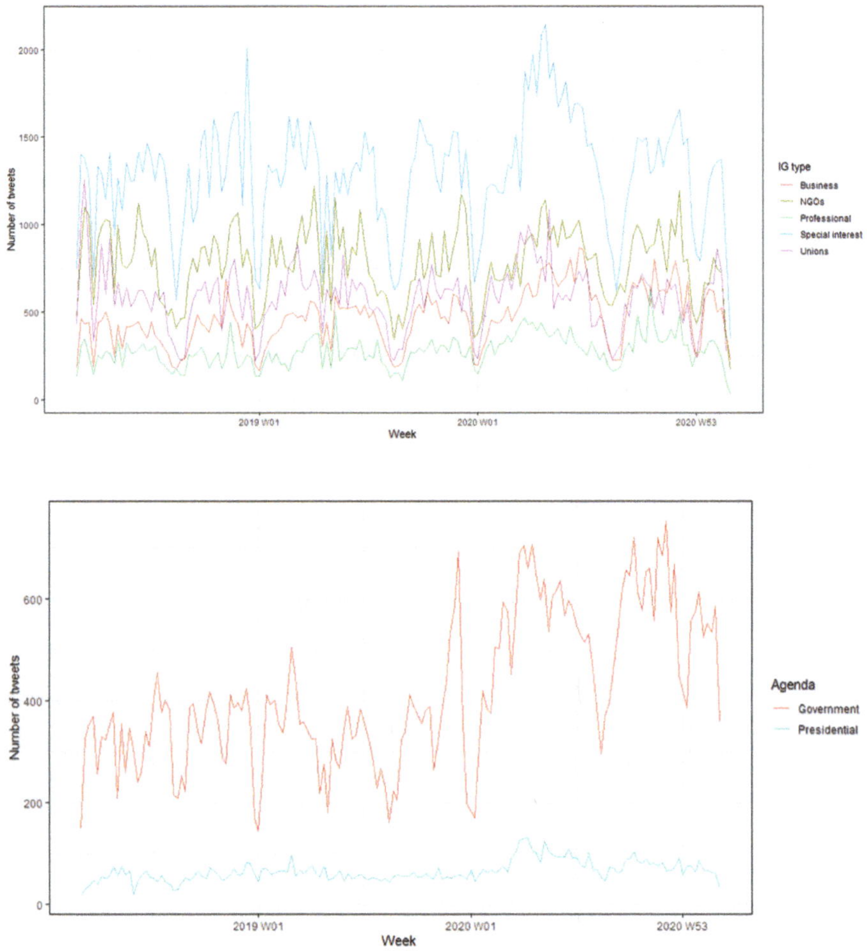

Figure A1: Number of Weekly Tweets by Interest Group Type and Government and Presidential Agendas

CHAPTER

7

Social Media as a Framework for Predicting and Controlling Social Protest in the 21st Century

Santiago Giraldo-Luque
Autonomous University of Barcelona, Carrer de la Vinya, Faculty of Communication, Campus UAB, 08193, Bellaterra, Barcelona (Spain)

Introduction

Since the characterisation of the internet and social media networks as scenarios of open and democratic participation (McLuhan 1969, Wiener 1969, Bennett and Segerberg 2012, Castells 2012), their study as catalysts of social mobilisation has been a recurrent one. The motivation for the numerous studies, which appeared mainly in the second decade of the 21st century, was generated above all by the emotional impulse of the Arab Spring. Ten years later it has proved to be a not very encouraging experience for democracies in some countries, and far removed from all the expectations of change generated on the web in the spring of 2011. The cry "the revolution will be retweeted", characteristic of the streets and networks of those 'springs' that swept the world, has gradually faded away.

The new forms of organising social protests mediated by social media networks (Ortiz 2016) have therefore been the subject of various analyses. On the one hand, the emotional contagion produced in a social network has been studied. Emotional contagion in networks, according to several authors who have conducted experiments on the manipulation of messages that users receive on their social media (Bond et al. 2012, Coviello et al. 2014, Jones et al. 2017) can lead to physical social mobilisation.

On the other hand, differences have been identified between traditional protests and mobilisations that involve a high use of social networks (Anduiza

Email: santiago.giraldo@uab.cat

et al. 2014), which has consolidated the concept of "connective action". The concept, coined by Bennett and Segerberg (2012), proposes two characteristics that are fostered by the use of digital communication and social media. Firstly, political content takes the form of general frameworks that can be personalised and adapted to different contexts (Anduiza et al. 2014), a process that implies the loss of institutional political centrality (Bentivegna, 2006) without the need for individual or group adherence or commitment to a prior ideological organisation (Cammaerts and Van Audenhove 2005).

The second characteristic highlights the possibility for citizens to share cognitive resources through trusted social media networks, removing the need to resort to traditional institutions. Anduiza et al. (2014), Eltantawy and Wiest (2011) and Castells (2012) also place a high expectation and prominence on social media as they can guarantee a mobilisation process and ensure an active and participatory role of citizens in collective action.

The compilation by Ortiz (2016) also describes how studies on the new wave of social cyber-movements have mainly dealt with the transformations of communicative resources and repertoires of action, the emergence of a new type of movement in the network era, and reviews on the theoretical framework of social mobilisation.

Despite studies categorising social media and the internet as indispensable arenas for political change, and as tools that facilitate participation (Castells 2012), the actual outcome of protest actions channelled through the social media is not evident (Torrego and Gutiérrez 2016) and in most cases examples of citizen protests have died out for two main reasons. Firstly, because political systems have reacted adequately to the challenge of protest (Lynch et al. 2017). In this case, mobilisations have been repressed, and stifled with some minor measures taken, or absorbed by systemic institutionality (Giraldo-Luque 2018).

Secondly, the citizen actions have been reconverted into institutionalised political movements (Casero-Ripollés et al. 2016) that strengthen the legitimacy and the discourse of systemic openness of the institutionality that, in principle, they rejected. The key to the success of the social mobilisation that emerged between 2011–2012 consisted in the maintenance, via association and the construction of ties, of the established protest. A requirement that meant the articulation of a broad and active popular base under a common understanding (Chomsky 2011). But this condition now proves impossible except in smaller organised groups or communities that have accumulated previous work. Such characteristics of contemporary mobilisation begin to outline a critique of discourses such as that of Cammaerts and Van Audenhove (2005), when they point out as distinctive of the new mobilisations, a structuring from the use of the internet, even without having a prior link to a collective organisation that weaves an identity (Tufte 2015).

Ten years after the construction of the collective emotion of social mobilisation, guided by social media, different political events (also channelled through social media) have mitigated the impact and the trust placed in the democratic and open new channels of communication. The concentration of information and communication power in a few platforms; the lack of control, or poorly exercised control, of public interest information flows managed by non-transparent private algorithms; the sale of user information carefully segmented with the intention of political manipulation to specific actors and for electoral purposes; censorship established under private logics and without any political, democratic or citizen control; the emergence of the phenomena of polarisation, affective polarisation and fake news, and the increasingly powerful capacity of social media networks for prediction and individual and social surveillance, means that the former allies of collective mobilisation are now viewed with a critical eye.

In this context, the chapter proposes a critical theoretical reading of the use of social media as the main space for communication and articulation of objectives of contemporary social mobilisations. They are identified by allocating a significant part of their resources to communication mediated by social media (Feenstra et al. 2016). The text reflects on the narrow view of the democratic and participatory process enacted by social media, while warning about how dominant companies in platform capitalism can dominate, predict, condition and control the spirit of contemporary social mobilisation (Fernández-Rovira and Giraldo-Luque 2021). The clearest example of the above positioning is that there is no social mobilisation in the 21st century, outside of social media, that opposes the monopoly of information and social exploitation, which is formed by the GAFAM group (Google, Amazon, Facebook, Apple and Microsoft), and some Chinese companies, such as Tencent, Alibaba, ByteDance or Sina.

The first part of the chapter presents the theoretical scaffolding on contemporary social mobilisation, in which the concept of social mobilisation is presented and in which the theoretical method of approaching the analysis of social movements in the 21st century is constructed. In the second part of the chapter, using the previous theoretical elements, a reading is made of contemporary social mobilisation delineated by the use and channelling of social media. Based on the application of the analytical model, the text criticises the absence of strategies, the centrality of actions on the handling, management, and search for trending topics on social media, and the lack of temporal permanence of mobilisations due to the absence of a collective identity. The last part of the chapter offers some final thoughts on the complex relationship between social mobilisation and social media.

The chapter emphasises the power of social media as guarantors of the current systemic conditions, which are largely favourable to their interests. With their power to concentrate on individual and collective information, and with the capacity to process an infinite source of data (the interactions of users with their digital platforms), social media not only define the themes of public discussion but can also guarantee, or deny, the visibility of a social mobilisation that promotes the defence of a human right. This is the power that citizens have unwittingly given to social media.

A Theoretical Method for Approaching the Analysis of Social Movements

The conceptual approach to social mobilisation and its study in contemporary society is structured around three conceptual proposals. Firstly, the most classical concept and the phases of the mobilisation cycle proposed by Sidney Tarrow (2016) are presented. Secondly, the definition and characteristics of mobilisation developed by Rocío Ortiz (2016) are taken up again, in which traditional definitions are updated and an approach to social cyber-movements is made. Finally, the three-dimensional analysis proposal by Santiago Giraldo-Luque (2018) is introduced. It outlines the analysis of social mobilisation through the categories of strategy, action, and objectives, and introduces the levels of moral development applied to collective social behaviour.

Sydney Tarrow defines four basic elements of analysis for the study of social movements: 1. Opportunity and cycles. 2. Phases of opportunity and constraint. 3. Innovations in cycles. 4. Dynamic elements of cycles of confrontation (2016). For the author, the first phase of opportunities and cycles begins with the articulation of concrete demands that can challenge the traditional positions of established interest groups in regional or national power (2016) and that can lead to the generation of territorial coalitions that shape social organisation. These are specific demands of a given group, which highlight the vulnerability of elites, question the interests of other groups, or point to possible convergences that facilitate coalitions.

Tarrow (2016) defines the opportunity and cycle phase as fuelling five specific communication processes: information flows more quickly; the flow of information about the movement's actions and demands increases; the publicised display of the determination of social discontent grows; there is greater political attention from institutions; and the frequency of interaction between groups of dissidents and between them and the authorities increases. The phase of opportunity and constraint describes, according to Tarrow (2016), the opening of political options for consolidating demands to the pioneers of the movement. Opportunities do not work the same for everyone throughout the cycle, as the options that open for pioneers may not be available to supporters. Instead, pioneers may in turn create opportunities for others.

Innovation in cycles evidences the invention of identities, tactics and demands as the social movement evolves (Tarrow 2016), as well as the importance of the generation of symbols projected into society by the movement, frameworks of meaning and new or renewed ideologies to justify collective action. This can then be incorporated into the political culture in a more diffuse way and be a source of inspiration for future movements in a less militant way.

Finally, the phase of the dynamic elements of cycles of confrontation is developed by the author based on three causal mechanisms: diffusion, exhaustion and radicalisation/institutionalisation. About the diffusion of conflict, Tarrow

(2016) indicates as a key element of protest cycles the extension of the proclivity for collective action for both unrelated and antagonistic groups, within what the author qualifies as the expansionary effect of collective action. Moreover, according to Tarrow, it is not only about diffusion, but also about the transfer of protest to system levels where new opponents, potential allies and different institutional contexts mark its progress.

As for the exhaustion of protests, Tarrow indicates that it translates into declining participation and uneven neglect across movement sectors in both the centre and the periphery. On radicalisation/institutionalisation, opposing processes, which can occur simultaneously, Tarrow explains that radicalisation occurs when there is a slippage towards extremes of ideological commitments and/or the adoption of more violent forms of protest. Conversely, institutionalisation occurs when there is a move away from extreme ideologies and/or the adoption of more conventional forms of protest.

Ortiz (2016), for his part, proposes nine characteristics of social movements that differentiate them from other collective phenomena generally associated with social protest or collective action. Firstly, he defines a social movement as a collective actor, that is, as "a group of people acting in a social environment" (Ortiz, 2016). Ortiz also points out as a second element that the objectives of mobilisation must be directly related to promoting or resisting social change. This is a characteristic that the author associates with Laraña's (1999) conception of the solidarity that can build a mobilisation to promote – or prevent – social change.

Thirdly, according to Ortiz, social movements set public goals that aim to solve the social problems of a whole community and not only of the movement's members. Thus, the author explains the fourth characteristic: "the development of public goals is motivated by the existence of a conflict" (Ortiz 2016). A conflict that can be described, according to Ortiz, in the perspectives of Touraine (1969) in identifying an adversary to which the mobilisation is opposed, and by Tarrow (2016), for whom collective defiance is a hallmark of social mobilisation. The very identification of the conflict, its actors, the collective challenge, and an adversary also promotes the fifth characteristic: the construction of a collective identity that defines the meaning of its action.

The sixth characteristic proposed by Ortiz is the intentionality of the movement's actions. The author takes up the theories of Castells (1997) to define mobilisation as "conscious collective actions whose impact [...] transforms the values and institutions of society" (Castells 1997 in Ortiz 2016). The characteristics of identity, conflict, promotion of social change and structuring of public objectives determine the character of continuity as the seventh element in the analysis of social mobilisation.

Horizontal reticularity, "which implies flexibility in the relationships between group members and reflects a horizontal structure in decision-making" (Ortiz 2016), defines the eighth characteristic described by Ortiz. Finally, the ninth defining feature for the analysis of social movements is their civil society-oriented communicative intentionality. Ortiz highlights that "the actions of these collectives do not take place in the institutional sphere, but, through communication strategies, they try to influence civil society and public opinion" (Ortiz 2016).

Finally, Giraldo-Luque's (2018) proposal introduces a three-dimensional analysis of social mobilisations based on the dimensions of strategy, action, and objectives. Strategy is assumed as a "correspondence between the actions chosen as an expression of protest or demonstration and the democratic evolution of the political system in which the action takes place" (Giraldo-Luque 2018). The author focuses on the adaptive actions of mobilisation previously pointed out by studies such as those of Anduiza et al. (2014) or Castells (2012) and in the framework of the duality emotion-reason that guides the contagion or adhesion of participation in collective social action. Giraldo-Luque defines strategy as based on the understanding and appropriation of one's own context and the capacity of the actions chosen in the same context to generate challenges to the political system (Giraldo-Luque 2018).

The second element of analysis, action, is described by Giraldo-Luque as "the type of goals that are set by the collective and as the capacity of the protest to achieve its goals without requiring or demanding an institutional response" (Giraldo-Luque 2018). The author takes up some of Bentivegna's (2006) and Ortiz's (2016a) previous postulates by seeking to distance the actions of social movements from established centres of institutional power. At the same time, he raises the question of the type of institutional response. The character, systemic or innovative, of the institutional response "also makes it possible to outline the scope of the goals defined by the mobilisation" (Giraldo-Luque 2018).

Finally, Giraldo-Luque explains that the objective of a social movement must meet two basic conditions. The first is that it must map out its actions (including its repertoire) in the long term. The second is that the definition of its goals must be identified with a problem associated with the guarantee and exercise of fundamental human rights. For the author, these two conditions guarantee a "coherent, diversified, autonomous and dynamic line of action represented in concrete actions" (Giraldo-Luque 2018). Giraldo-Luque's proposal also introduces the distance of the social movement from previous traditional organisations, pointed out by Cammaerts and Van Audenhove (2005), as well as the creation of a collective identity (Touraine 2005) that also ensures the longitudinal projection of the movement's own actions (Laraña 1999, Ortiz 2016).

Giraldo-Luque's proposal is based on the stages of moral development (Kohlberg 1992, Habermas 1981) applied to collective social behaviour (Alutiz 2004, 2010). In the post-conventional stage of social mobilisation, the definition of a social movement is developed, as well as a characterisation of the use of communication and social media within the elements defined as strategy, action, and objectives. In the post-conventional stage, the strategy of a social cyber-movement finds in communication a suitable tool to convene and disseminate its actions (Alonso-Muñoz, and Casero-Ripollés 2016). In general, though, they have low expectations in specific mechanisms, such as social media, as they fail to articulate discursive consensus and are "integrated into a rational digital communication strategy" (Giraldo-Luque 2018). Moreover, it is oriented as a discursive unit (Lynch et al. 2016). For its part, the dimension of action integrates the communication framework as an additional pressure factor that "accompanies the *de facto* actions carried out by the movement to achieve social legitimacy" (Giraldo-Luque 2018). From this perspective, social media are tools for other types of strategic actions (Feenstra et al. 2016). Regarding the objectives, communication describes a long-term and consensual discursive structuring process, which guides a broad repertoire of autonomous protest actions (Giraldo-Luque 2015).

According to Giraldo-Luque (2015, 2018), the social movement emerges as a collective construction that places its objective in the long term and is oriented towards the conquest of rights. In its consolidation, the construction of consensus on actions, guided by the moral perspective of justice, allows it to outline a strategy that generates political challenges and, at the same time, to propose specific actions assumed from the perspective of autonomy.

The theoretical and conceptual proposal is a qualitative approach to the phenomenon of social mobilisation represented mainly in the social mobilisation of the second decade of the 21st century. The method employed aims to critically analyse some of the actions of social movements, in terms of the use and constraints imposed by social media, and in the light of the theoretical categories proposed by Tarrow (2016), Ortiz (2016) and Giraldo-Luque (2015, 2018). These theoretical categories are presented as a methodological operationalisation of analysis.

The chapter approaches contemporary social movements based on Tarrow's four phases of the protest cycle in which, in each case, the categories and characteristics proposed by Ortiz and Giraldo-Luque are linked and grouped into three dimensions. The framework of analysis can therefore be represented in Figure 7.1.

Each of Tarrow's four processes will be explained based on the characteristics described by Ortiz (in italics) and Giraldo-Luque (in bold). The grouping of Ortiz's and Giraldo-Luque's characteristics is proposed as a synthesis of the elements that the two authors have in common.

Analysing Social Mobilisation in the Age of Social Media

Opportunities and Cycles: The Impossible Consensus

One of the main problems with social mobilisation in the age of social media is that it is not capable of generating political opportunities for the whole of a society and a citizenry that are absolutely individualised (Marcuse 1987). Some of the experiences of social mobilisation prior to Instagram and Twitter were able to generate opportunities by building consensus on different issues. Social mobilisation managed to articulate a series of specific demands (Tarrow 2016)

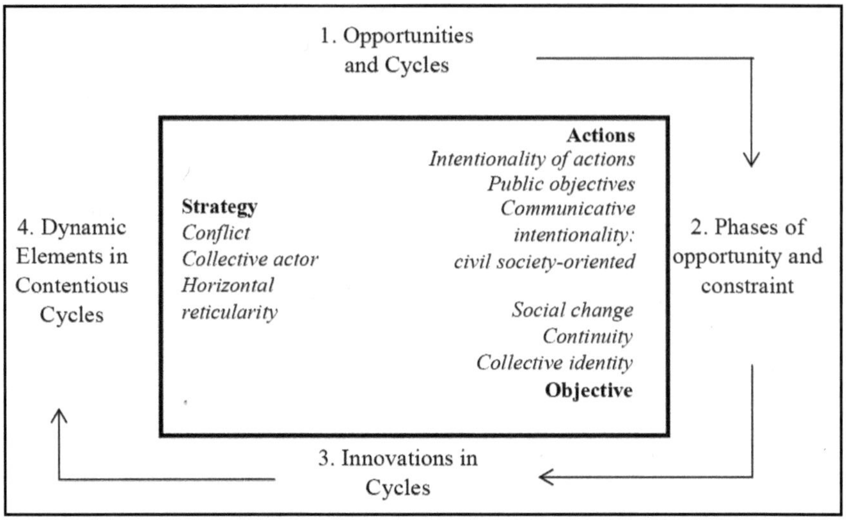

Figure 7.1: Analytical Theoretical Framework

that pointed out and took advantage of the poor response (or even the exclusion) of institutions when faced with the satisfaction of certain rights of the affected groups.

In the absence of collective articulation of a consensual demand, the possibility of constructing an intentionality of action is blurred in social media (Ortiz 2016). On the other hand, the logic of demand, individualised, registered, monitored, and controlled by private spaces of conversation – converted into a database – follows the traditional path of demanding a solution from traditional institutions (Bentivegna 2006, Giraldo-Luque 2018). In its conversion to the 21st century, the demand can have the discursive and intersubjective depth of 280 characters.

Hundreds of billions of meaningless, unarticulated and, logically, unsuccessful demands are posted on social media every day. Likewise, the success of the demand is not measured in the satisfaction of the need expressed collectively, but in the capacity to generate an echo in the same social media network. Thus, if the demand achieves the necessary success, such as a trending topic, some form of institutionality (public or private) may respond individually to the demand, satisfying it. But the response is made to the individual who has generated the demand. It is an individualised response. It does not respond to an articulated collective and, therefore, does not resolve the social problem that affects the non-fulfilment of a right. Similarly, the individualised demand, contrary to what Tarrow (2016) proposes, does not question the interests of dominant or traditional classes and which constitutes a dispute with the interests of citizens organised as a social mobilisation. The demand generated, without the articulation of the collective consensus of an affected community, does not generate convergence on a medium or long-term social, political or cultural project.

The absence of collective demands, as a consensus, weakens the structuring of social movements. On the one hand, the objectives are not related to the promotion of social change. Likewise, without objectives, without consensus results that aim to solve the problems of a community, social mobilisation in social media networks disappears to privilege the individual (Ortiz 2016). On the other hand, within the individual, social media only rewards those who benefit from having high levels of public notoriety in the sphere of vanities of the platforms in which the system of asymmetries of visibility power governs (Fuchs 2014).

Another element that determines the possibility of opening political opportunities for social movements refers to the five communication processes mentioned by Tarrow (2016). Although in social media networks the flow of communication that generates an action of social mobilisation can increase rapidly and exponentially, while it is possible to express a greater determination of social discontent, the characteristics related to the degree of political attention from the authorities are not developed, nor is the frequency of interaction between different actors and groups with propositional or constructive intentions increased. The logic of the social media network, in its capacity to impact a population that is limited to the set of users' contacts, generates at least five problematic situations for social mobilisation.

Firstly, it reduces other types of relevant communicative actions because the intention or strategy of communication is reduced to social media as the only possible medium. Secondly, it assumes that the communication of collective actions, as well as expressions and opinions of solidarity on social media, is an act that is sufficient for citizen political participation. Thirdly, it uses a channel, the only channel, in which billions of instant messages compete for attention and in which the construction of political discourse and social organisation competes with sexualised swimming costume photographs, biscuit recipes, rap and reggaeton videos, memes, fake news, and summaries of sports competitions. Fourthly, the space of mobilisation, their online profiles, become epicentres of political polarisation and hate speech that make it even more difficult to build social consensus. Fifthly, communication on social media, that is, the entire communication campaign, is limited to short, reduced, simple promotion of their actions and demands, adapted to the formats defined by the platform, as it is impossible to build consensus on the demand in this online communication strategy.

A social or collective action, which does not start from a prior social consensus, and which only communicates demands or actions, only reflects symbolic emotive actions easily controllable over time, easily assimilated by the political system. This is what happened with all the social mobilisations that emerged in the northern and southern Mediterranean at the beginning of the second decade of the 21st century.

Given the omnipresence of social media in society, communication through social media networks is not conceived as a strategy but becomes the movement itself. Social movements within controlled networks are not relevant and collective action becomes predictable, calculated, with little power of communicative agency (Ortiz 2016). Nor does communication become a pressure factor that "accompanies the *de facto* actions carried out by the movement to achieve social legitimacy" (Giraldo-Luque 2018).

In the cycle of mobilisation, or of the expression of protest in social media, the vast majority of individual or collective actions that express social demands end up being assimilated, silenced, in the infinitude of the feed cycles of the structure of the functioning of social media.

Social media networks fulfil, socially and politically, a channelling function of all communication structures. They act as catalysts and neutralisers, assimilators (Tarrow 2016), of social mobilisation: they function as a form of social self-regulation, monitored (Foucault 1979) and supervised by an artificial intelligence system, which is also private. In the social media network, the restrictions for the granting of demands and for the opening of new opportunities for autonomous collective actions are not diminished (Giraldo-Luque 2018). On the contrary, the illusion is created that the channel for their, individualised, expression is opened up, thus guaranteeing an optimistic and regulatory vision of the social sphere in the face of the apparent openness of contemporary democratic political systems.

Phases of Opportunity and Constraint: The Era of Individualisation

While the demands and actions of collective mobilisation in the 20th century allowed the extension of rights to marginalised or excluded groups (Ortiz 2016), the universe of social media determines the physical disarticulation of social mobilisation with common objectives defined in organised collective deliberation (Giraldo-Luque 2015). As noted above, the satisfaction of rights through the manifestation of nonconformity in the digital scenario is carried out in an individualised manner and benefits, as Tarrow indicates, only the pioneers of the demand and never those who join later for an individualised benefit (2016).

The individualisation of social mobilisation destroys the conception of both social change and the presence of a conflict between collective actors, which determines the definition of social movements (Ortiz 2016). An individualised solution does not solve community problems, but it does achieve the systemic silence of those who managed to raise their voices amidst so much noise. Social media networks also act as a vehicle for injustice and as a "democratic" channel that creates the illusion that individualised citizen actors can be heard. In the absence of consensus, conflict between actors does not exist either. Its definition is erased by the individual demand.

The universe of complete individualisation promoted by social media platforms undermines society's capacity to articulate collective demands and privileges solitary action, associated with its impact on the network user's own social ego. Action in the network is not conceived as individual, with the intention of extending rights to the collective, but as a possibility of receiving an individual benefit. At the same time, the demand is made from an institutionalised channel (within the framework of a controlled and regulated, algorithmic system) and receives, in its functioning as a systemic machine, an institutional response, which limits the possibility of developing an autonomous (Giraldo-Luque 2018) or post-conventional (Habermas 1981) resolution proposal.

Social media determines the conduct of social mobilisation and characterises its discourse. In this sense, they are the ones that create the context in which the mobilisation takes place and in which its story, its narrative, must be framed. Thus, social mobilisation ends up by carrying out actions that must be framed and projected onto the very context or communication system constructed in a controlled manner by the social network platforms. The intention of the action, sometimes even violent, is that it be retransmitted via Twitter, Instagram or Tik-Tok, but the coherence, as a strategy (Giraldo-Luque 2018), between their own actions and the framework of the political-institutional system in which they are carried out is never considered. On the other hand, the limited diversification of protest actions (Anduiza et al. 2014) linked to mobilisation turns their actors and their actions, controlled by internet networks within a conventional system of action, into systemic elements that legitimise the established order.

Innovation in Cycles: Homogenisation and Loss of Rationality of Mobilisation

Social movements in the new century have lost the capacity to innovate in the repertoire of their protest actions. They are not creative and the degree of spontaneity, only a little linked to the solid discursive construction of a project in defence of rights, has made social media their main place of collective and individual expression, simplifying their message as much as possible. It is difficult for social mobilisation actions to be reflected in mass communication scenarios that go beyond the limited reach of 99% of the messages that circulate on social media networks.

The cultural and political homogenisation that occurs in internet networks also limits the construction of solid collective identities. The absence of innovation in actions, linked to reduced spaces of conversation and controlled in the dispute through automatic algorithmic techniques, the automation or standardisation of common sense (Celis 2015), homogenisation, slows down any real impact of the impulses of individual or collective mobilisation. Focused, articulated, channelled, and intertwined in the interactions of the dominant communication platforms, actions are born and die in the predictable and easily censurable space of Twitter, Facebook, or Instagram. Public political dialogue, which takes place in the same private space of the social media platform, also competes for pre-eminence within the intense information market and denies representativeness or interest in apparently mobilised, but non-consensual, demands. Competition, among all the options, ends up blurring rationality. It rewards advertising and emotive discourse, under the limiting and constrained logic of the platform.

According to Tarrow (2016) innovation in protest actions involves the invention of identities, tactics and demands, an impossible path when mobilisation actions are concentrated or channelled through a single communication space.

The spontaneity of the collective reaction that governs the logic of behaviour in the social media network contradicts the principle of planning the common action of a mobilisation, a characteristic that guarantees the temporal continuity of the social movement. At the same time, it is impossible to build a solid identification, a "coherent, diversified, autonomous, dynamic and represented in concrete actions" (Giraldo-Luque 2018), within a context of social change (Ortiz 2016).

In the construction of a collective action in the network, which is guided by homogeneous emotion and not by the discursive articulation of a social need, previously discussed, there is no adequate degree of organisation for its strategic action, which is only possible with the construction of a collective identity. The expression of social mobilisation in the net is only structured around an advertising slogan and there is no identity of its own (Giraldo-Luque 2018) that acts as a symbolic reinforcement of the construction of collectivity (Touraine 2005) and defines the meaning of the action (Ortiz 2016).

Thus, although there are nodes that send messages to each other and interact, mainly through the same channels of the dominant platforms, the actions are not established under the figures of "horizontal reticularity" (Ortiz 2016). This typical construction, as a characteristic of social cyber-mobilisations, guarantees the participation of all actors. But on digital platforms, different actors have dissimilar degrees of visibility (Fuchs 2014), and resources (cognitive, social, solidarity, cultural) are not shared, as was the case in the 20th century among well-established social networks of trust (Bennett and Segerberg 2012).

In the last characteristics of innovation in cycles (Tarrow 2016), social mobilisation actions that are born or structured in social networks also fail, for the most part, to involve a high level of citizen participation in the proposed actions; they do not manage to impact traditional or alternative media agendas on a massive scale and capture the public interest of society in general, and they do not capture the attention of political, economic, social or cultural elites. What is interesting is that a good part of the people who use social media consider that they do. It is the communicative power itself that sets the dominant ideological framework.

Likewise, current social movements are also unable to generate symbols as a mark of meanings projected to society (Tarrow 2016), and which give meaning to the continuity of a collective resistance proposal. What are the symbols of student social mobilisation today? Of the working class? Of the workers? Their demands are not very well known either.

Dynamic Elements of the Cycles of Confrontation: Exhaustion and Radicalisation

Tarrow (2016) describes some mechanisms within the cycle of mobilisation actions that can be applied to social movements linked to social media networks: diffusion, exhaustion, and radicalisation/institutionalisation. Within the network, the false imaginary that a message can spread to the whole of humanity is multiplied. This is the illusion that the individual lives with when he or she accepts the ideological background of social media networks. Dissemination, concentrated in a dominated and controlled network, is limited and the expansive effect of collective action is not realised. Posting a message on social media is not participation. For Tarrow, diffusion also incorporates elements in which a transfer of protest occurs at system levels where new opponents, potential allies and different institutional contexts mark its progress, a process that is undoubtedly impossible to realise in the social media network. What does happen, within the network and through its dynamics, is that processes of exhaustion and radicalisation are automatically triggered by diffusion.

In the first case, the emotionality inherent in the development of network actions, not rationally sustained, tends to fall over time. Experiments in measuring behaviour on networks with respect to a social mobilisation event (Giraldo-Luque et al. 2018) show the emotional transition from a very high peak of participation to an abandonment of the conversation and the expression of solidarity with the mobilisation, until its virtual disappearance or trivialisation within the infinite market of emotive informative news. The logic of the algorithm, in this case, linked to the control exercised by the large platforms, would even make it possible to measure the degree of impact of a protest or social mobilisation and how long it remains active.

In the case of radicalisation, as pointed out by different studies (Lewis 2018, 2019, Horta et al. 2020, Munn, 2019), social media networks promote affective polarisation and produce the effect of multiplying discursive emotionality, while denying the space for argumentative rationality. In fact, the latest studies on affective polarisation (Fernández-Rovira 2021, Carrillo 2021) indicate that when a message is published on social networks, the reaction can be of two types. In the first case, a response is generated that is completely contrary to the ideology of the message. In the second, the interaction is one of complete affiliation with its ideological content. But in both cases, as the interaction goes one step further (which defines and revitalises the two-step theory of communication (Katz and Lazarsfeld 1955), the level of passion in the discourse becomes increasingly intense, more affective and, therefore, less rational and argumentative. In short, more radicalised. Political discussion, defined as the construction of common sense exercised as the communication of a shared judgement on what a community defines as just or unjust (Celis 2019), becomes impossible when channelled through social media. Radicalisation events, as much as exhaustion, that happen very quickly in the dynamics of social networks, are the basic characteristics of

21st century social mobilisation channelled and controlled in the private spaces of Google, Facebook, Twitter or Tik-Tok.

Social Networks as a Systemic Element: The Revolution will not be *Instagrammed*

If protest and social mobilisation is concentrated and channelled through social media, it can never have the transformative effect inherent to social change that characterises social movements (Ortiz 2016, Giraldo-Luque 2018). When the dispute or the construction of discourse is concentrated in channels that deny argumentative inter-subjectivity, the only thing possible is the emotional (advertising) appeal that undermines the construction of collective identities. In the social network, the only thing that the emotional explosion does is to produce jumps of extremes, which become polarised and radicalised as soon as the interaction of messages deepens.

In this way, social movements that use the dominant digital communication channels become predictable and systemic. Social networks, as objects that monopolise spaces of dispute, mobilisation and communication, fulfil a fundamental systemic function for the maintenance and reproduction of current political systems. Through their actions, which have sharpened the capitalist accumulation structure of global society, they have ensured that protest remains in controlled circuits, under conventional structures (Giraldo-Luque 2018) and thus prevents social discontent from overflowing the systemic limits to generate real challenges to legal, regulatory and economic systems. If mobilisation remains within the limits of the individual and under the communicational structures of social media platforms, protest actions will be easily assimilated from the private space of algorithmic control, or from the traditional structures of institutional power (media, political limitation/exclusion or inclusion, or legitimate and monopolistic exercise of force) (Giraldo-Luque 2015).

The End of Social Mobilisation

The latest wave of democratisation heralded by the transitions of some of the Arab countries to regimes with democratic characteristics, such as electoral processes in which different political options compete, implies the adaptation of almost all countries in the world to a system that is socially legitimised and protected by most citizens. The demand for democracy that filled the streets of many cities around the world in the spring of 2011 ratifies the widespread belief among the population that the democratic system can guarantee better living conditions in an environment of civil and political liberties and under a scenario in which, at least theoretically, all citizens can participate in the public decision-making process.

The basic approach that sustains democracy as a participatory system, through elections or other more complex forms of participation, gives a privileged status to the democratic political system. The mantle of freedom of choice, and of expression, covers democratic practices and reinforces the social contract that is established between citizens and democratic states. As signatories, citizens not only respect the pact, but also defend it against possible attacks. Their defence also involves safeguarding the conditions of the pact that affect individuals as a collective: rights, security, or social peace. These are important foundations for protecting democracy itself as a political system.

However, democracy in the 21st century has a major challenge centred on the return of the values with which democracy was born in classical Greece. It is the real possibility of citizens' participation in an environment guided by representative democracy.

In the structuring of 21st century democracy, absence of the deliberative or discursive landscape from which it started as a political system against authoritarianism, the acceptance of the social contract includes an instinctive satisfaction in the system of non-freedom, or imagined freedom, that helps the same systemic, capitalist, democratic project to perpetuate itself (Fuchs 2017, Marcuse 1987). The rising standard of living, at least in industrial society, but declining in the Information age, in the post-crisis world of 2008, lent legitimacy to the democratic and authoritarian political process. In the third decade of the 21st century, apparent freedom has ceased to be rationalised through higher standards of living (Piketty 2014). Its rationalisation and internalisation, sophisticated domination, is produced through the psychological domination of the individual and his or her psychic satisfaction through the *like* button, through the endorphin loop produced in everyone when they receive a notification.

The impossibility of building a dialogue with people who, democratically, think differently, is the main characteristic of the use of social media as a means of communication and structuring of communication in social mobilisation. The absence of dialogue is a consequence of the same process of simplification of life, with an aggravating factor. The empowerment of discursive univocity is increasingly sophisticated given the characteristics of datification. Social media acts as sounding boards or "echo chambers" (Pariser 2017) of ideological preferences, and they cancel out the possibility of confronting ideas with other perspectives on life. The engines that set up predictions make sure that these kinds of uncomfortable behaviours, where we have to converse with someone who thinks differently, do not happen.

Democracy in the 21st century is going through its darkest period since the worst era of absolutism. The Greeks ensured public discussion just as the liberal revolutions – the English, American and French – gave their parliaments the basic task of acting as regulatory mechanisms in the face of centralised powers concentrated in a monarchical figurehead. In the renaissance of democracy after the World Wars of the 20th century, parliaments guaranteed democratic and representative public discussion. Twenty-first century parliaments, however, are a caricature of public deliberation. Discursive enunciations have abandoned the hemicycles for the impact that a *lucid* sentence can have in the next few minutes on Twitter, Instagram or Tik-Tok. After a while, it is time to look for a new witticism.

The central element of social mobilisation is neither the flow of information nor people's ability to express themselves. Both elements are important because without them, without good information and freedom of expression, it is impossible to build a dialogue between actors. But the basic point of a democratic process has historically been dialogue and deliberation. The new communication tools were envisaged as a new virtual agora in which everyone could participate. So far, we have only seen hysterical shouting, banal messages that reach millions of *likes* and armies of trolls ready to fill any online discussion with rubbish. Social mobilisation is impossible on the internet because there are no possibilities for deliberation that would force people to build consensus on actions, policies, and public life.

The very concept of life, interest, and public opinion conflicts with the dynamics of social media, as political issues are of much less interest than entertainment topics such as sports or show business (Fuchs 2014). In the concentrated market of information, the competition for attention always disregards, for most users, political discussion. It is traditionally removed from their everyday interests. As we have seen in several of the preceding chapters of this book that the algorithmic definition system of social media networks prevents citizens from accessing public agenda issues if they are not part of their daily and interested consumption, predefined by their browsing behaviour on the web.

The absolute connection to devices, to copper wires, as seen in the streets and means of transport of large and small cities around the world, saturates the capacity for real communication. Hyper-connectivity generates the sensory isolation of man from his environment and creates incommunication in micro-relationships as well as in political and democratic macro-relationships. In cyberspace there is no unifying global village. What exists is a normless bombardment of a multitude of messages representing inconsistent and incompatible universes (Žižek 2006).

In the same critical vein, Barber (2003) points out that modern, large-scale societies present several problems (isolation, civil alienation, declining trust) that cybersociety is unlikely to adequately remedy. Segmented media further deteriorates public space and prevents the construction of a meeting place. Their specialisation in market niches (individual beauty) offers an advantage for policy against public deliberation, as well as undermining conversation, thereby (for politicians) achieving the adequacy of a controlled common ground and the delimitation of citizen participation.

Inaction and Uniformity Networks

Today's democratic-totalitarian society demands the acceptance of principles and institutions that limit opposition and alternative policy debate within the established institutional conventional framework, within the *status quo*. The gradual reduction of direct political controls from the represented to the representative through associative and partisan mechanisms, abandoned in the transition from the industrial society to the information society (Giraldo-Luque and Fernández-Rovira 2021), is just another proof of the confidence in the effectiveness of technological control as an instrument of social domination. The new institutionalised political parties, some of which were former social movements, assume a minimalist programme and marginalise the radical social changes required by an *aliberal* society. Their struggle is contemporised by the rules of the parliamentary game (Marcuse 1987).

As Marcuse (1987) highlights, managed mobilisation through controlled technological apparatuses and networks makes the explosive and anti-social elements of the most human unconscious socially manageable and usable. It is dehumanisation. Power, exercised as a means of communication, is in charge of guaranteeing systemic freedom through the transmission of alternatives, individualised and gestated through datification, so that the individual *chooses freely*. It is the system, technified and communicative, which chooses the possibilities of individual development from sophisticatedly controlled environments.

While the 19th and 20th centuries saw a relationship of political and economic power based on the use of violence as a way of controlling the application of an established social contract, the Information Society maintains control under the paradigm of communication. It is through communication, its codes and languages, and the systemic capacity to assimilate conflict, to predict it, that power takes care of the harmonious integration of society. From a systemic perspective, also functional or ascribed to the technological universe, the reduction of the complexity present in the chaotic social, political, and cultural environment is achieved through the reduction – or prediction – of alternatives, thus making it finite, manageable. Within the scenario of the exercise of power, the systemic function of the communicative interface fulfils its mission by making the system itself live in the actors (Luhmann 1995). By being included within the system, or properly living within the system itself, the interactions between the different actors are kept under an environment of control while external challenges are assimilated. The relationship between the environment and the communicative system itself involves adaptation and change that responds to the challenges presented. The capacity to adapt avoids revolution or, in other words, the change inherent in the processes of social mobilisation: the revolution will not be *instagrammed.*

In communicative terms, the adaptations of the interfaces of social media networks, at the internal level, as well as the acquisition of multiple start-ups, in their external dimension, imply the logical adaptation to the challenges of a complex and dynamic world interacting with diverse and competitive systems. Targeted communication establishes meaningful links between actors and elements, giving them unity, while responding to the challenge of the environment. At the same time, it manages the necessary impulses to ensure the multiplicity of choices that determine the actions of the actors and determine the boundaries of the system. Communication builds a flexible and *free* fabric within the same systemic world or universe that ensures communicative feedback and, at the same time, the reproduction of the system. The guidance of choices, in the imagined freedom of the individual, is the ultimate expression of the exercise of power (Luhmann 1995).

Power as communication is reinforced in the orientation of personalised selections and in the construction of a shared, semi-universal code. The power relationship also implies the overcoming of violence prior to the exercise of power and exercises the memory of violence as a framework for legitimising its exercise of power. It creates a vision of expanded freedom that the individual assumes as truth and ends up defending it against the threat of past, or future, violence. Under the control of uncertainty, over security, or even over free time, the individual accepts the proposed, selected, predictive socialisation that the technological scenario offers. The human being, subjected to the uncertainty of necessity, assumes that the social technological system offers him/her only certainties. But the system does not eliminate uncertainties, it only manages them through the exercise of power and the management of insecurity, the fundamental focus of the political campaigns of, for example, Donald Trump in the United States, Jair Bolsonaro in Brazil, Vox in Spain, or Uribism in Colombia, which assumed the management of fear as a key point to maintain the highest possible level of attention.

The focus of the systemic perspective is acquired in the lessening of contrasts, in the declaration of the end of the conflict between the given and what is possible, between the needs satisfied and the needs to be satisfied. In assimilation, in equalisation, the levelling out of class distinctions takes place, which fulfils an ideological and cultural function. Their integration indicates not the actual disappearance of social classes but the preservation of the needs and satisfactions of the established system which are now shared by the underlying population. Men and women end up recognising each other not only in their commodities (a car, a mobile phone), but also in their capacity to influence as positioners of these needs and satisfactions (Marcuse 1987).

The one-dimensional (Marcuse 1987), systemic (Luhmann 1995) and communicative (Fuchs 2020) society succeeds in conquering transcendence, in controlling it, through institutionalised desublimation. The reduction-absorption of opposition establishes a qualitative difference in the world of political life. Culture, as much as politics, is instinctively reduced and regulated.

In the now classic terms of the 1960s, in which youth emerged for the only time as a historical subject, the configuration of the social movement as a subject implied the construction of a living body that was nourished by the awareness of its own condition as a social object. The individual recognised him/herself in a process of social interaction, deliberative and vigorous. The streets and the political dialogue, as a public space, allowed the encounter of common situations and the actors found innovative dynamics of social protest with the aim of claiming the rights promised by a democratic society in the process of construction.

The politicised society passed in a few decades from consciousness to stillness and its level of collectivisation, community building and social dialogue gradually fragmented. The social status of the subject has been individualised to the maximum through mechanisms of communicative social control. The great technological leaps have meant a fundamental qualitative change: the disarticulation of social relations, even family relations. The dialogical capacity of the social world to construct a joint project that protects its own rights has been lost. It is the loss of the human capacity to rebuild a dynamic and structured social fabric. It is the loss of the human capacity to converse (Turkle 2019).

The caricature of 21st century social mobilisation is produced solely through private networks that systematically controls public discussions and the states of opinion, moods, and happiness of much of the world. Social chaos is experienced when screens are suddenly switched off. To live without Wi-Fi is the negation of existence. It is to disappear. While in the international imaginary the Egyptian 'revolution' of 2011 is assumed as a victory of the citizenry over a totalitarian regime, the return of the military to power has meant, for example, the expulsion of foreign correspondents, the prosecution of opposition bloggers, the militarisation of universities and the closure of political competition. In Spain, 15M, which articulated a significant mass of indignation, faded into small, unstructured groups with no apparent consensual north. Its most obvious result, a political party, now plays in the institutional political chessboard as one more fragmented and concentrated actor that legitimises the systemic opening. Student mobilisations, motivated by demands that are not entirely theirs, as if they had no autonomy as young people, resort to 20th century protest systems. Stuck in the past, they ask teachers not to teach so that at least then the classrooms will be empty of students. They are not even able to convince their classmates of their common demands as students.

The mass that is mobilised, unconsciously, automatically responds to what private social networks consider fair, the definition of collective common sense, and turn into a political agenda or trending topic. The internet, like all its media predecessors, has not brought democratisation to 21st century societies. Cyber-mobilisations are volatile, apolitical, de-ideologised, unconscious. They function as a mass and not as a collective subject. Systemic processes are found in the demonstrations of mass groups in the streets, actions that legitimise the *freedom* of the system to openly express different points of view.

The theories of digital participation were put forward at the beginning of the second decade of the millennium either from the utopia of the interconnected and hyper-intelligent global village, of mass self-communication (Castells 2009) or from the consolidation of digital divides, and cognitive divides, which, in political matters, would guarantee the voting of the public from time to time to reproduce the conditions of a political class dominated by private interests. With the social mobilisation of the beginning of the second decade of the 21st century, periodic elections in many parts of the world have demonstrated the communicative power of the new media to harness the attention economy and generate the fear necessary for dehumanised ideas to govern millions of people. Fear has been taken over, as much as anger and animosity against the other. In indignation against the subaltern (the migrant, the woman, the black, the poor), there is mobilisation. In individual identification with a superior social class, which oppresses the same individual, there is mobilisation. We have taken on the discourse, needs and ideas of a class to which we do not belong. The biggest paradox is that we defend them with passion.

For the moment, networks and information highways can only be seen as placebos of systems that are neither democratic, nor free, nor user-controllable. The imaginary built on the internet as an element that can deepen the roots of democracy has been shown to be false. But not only that, social networks have been denounced by hundreds of experts who were part of the very creative heart of the monopolistic communicative system of the information society, as the most dangerous tool for the exercise of control over human life, thought and actions.

Self-constituted media power legitimises itself through *open* digital channels of expression, and its apparent information transparency validates its obscure and self-interested, uncontrolled practice. Social mobilisation, channelled through networks, is blurred by the systemic plasticity of demands and cyber-mobilisations, controlled by classic devices. It has been a long time since social protest actions, self-controlled and within democratic society, have generated significant systemic challenges to democracy. Because they all circulate through the cells of the nervous system of digital oligopolistic power: the social media networks.

The space for social dialogue requires the construction of subjects rather than the mobilisation of thousands of people in public squares. A social consensus is more valid than a trending topic. The only way to generate new spaces for the effective participation of organised citizens is using information and communication technologies as tools, not as objectives, to generate spaces for social discussion and, through deliberative actions, to produce awareness and social solidarity in the face of human rights such as work, life, freedom, health, education, housing, free expression and democracy.

Conclusion: The Hijacking of Social Mobilisation

Our minds can be hijacked: the tech insiders who fear a smartphone dystopia. This is the title of an article published by Paul Lewis in The Guardian in October 2017, in which important former employees of Facebook and Google intervene. In the text, James Williams, a former Google strategist, indicates that the fake news circulating on social networks, the bots created on Twitter or the efforts to measure target audiences milli-metrically based on data, are symptoms of a much deeper problem.

Williams points out that the problem is not just that special interest groups exploit the internet and social media to manipulate public opinion. The big problem, he argues, is that social media allows phenomena like Donald Trump or Jair Bolsonaro to secure voter empathy by capturing and retaining the attention of both supporters and critics, without bothering to create or manage outrage as a sentiment that promotes political action.

For Williams, Donald Trump's victory implied a point of no return: "the new, digitally supercharged dynamics of the attention economy have finally crossed a threshold and become manifest in the political realm". The emotional factory of outrage, according to Williams, not only distorts the way we view politics, but, over time, can change the way we think by making people less rational and impulsive: "We've habituated ourselves into a perpetual cognitive style of outrage, by internalising the dynamics of the medium", of the frayed social web. Williams argues that the fixation in recent years with George Orwell's fictional surveillance state, explained in *1984*, may be displaced. For him, the perfect metaphor was constructed by Aldous Huxley, author of *Brave New World.* Williams warned that Orwellian-style coercion was less a threat to democracy than a more subtle power of psychological manipulation, accompanied by "man's almost infinite appetite for distractions".

To the irrelevant state of political discussion that the current democratic framework entails, Williams adds the narcotic state of individual symbolism that dominates the brave new world of social media. "If the attention economy erodes our ability to remember, to reason, to make decisions for ourselves – faculties that are essential to self-governance – what hope is there for democracy itself?

"The dynamics of the attention economy are structurally set up to undermine the human will", Williams highlights. "If politics is an expression of our human will, on individual and collective levels, then the attention economy is directly undermining the assumptions that democracy rests on". Paul Lewis, in the interview, finally asks Williams:

> *"If Apple, Facebook, Google, Twitter, Instagram and Snapchat are gradually chipping away at our ability to control our own minds, could there come a point, I ask, at which democracy no longer functions?"*

Will we be able to recognise it, if and when it happens? Williams replies. And if we can't, then how do we know it hasn't happened already?"

References

Alonso-Muñoz, L. and A. Casero-Ripollés. 2016. The influence of social change discourse on the media agenda. The case of the Plataforma de Afectados por la Hipoteca. *OBETS. Revista de Ciencias Sociales*, 11(1), 25–51. doi:10.14198/OBETS2016.11.1.02

Alútiz, J.C. 2004. The moral stages of social evolution. *Journal of Sociology*, 74, 11-44.

Alútiz, J.C. 2010. An introduction to postconventional morality from sociology and political philosophy. *Postconventional*, 1, 8–19.

Anduiza, E., C. Cristancho and J.L. Sabucedo. 2014. Mobilization through online social networks: The political protest of the indignados in Spain. *Information, Communication & Society*, 17(6), 750–764.

Barber, B.R. 2003. Which technology and which democracy. *In*: Jenkins, H. and Thorburn, D. (Eds.), *Democracy and New Media* (pp. 33–48). MIT Press.

Bennett, W.L. and A. Segerberg. 2012. The logic of connective action. *Information, Communication & Society*, 15(5), 739–768.

Bentivegna, S. 2006. Rethinking politics in the worlds of ICT. *European Journal of Communication*, 21(3), 331–343.

Bond, R.M., C.J. Fariss, J.J. Jones, A.D.I. Kramer, C. Marlow, J.F. Settle and J.F. Fowler. 2012. A 61-million-person experiment in social influence and political mobilization. *Nature*, 489, 295–298. Doi: 10.1038/nature11421.

Cammaerts, B. and L. Van Audenhove. 2005. Online political debate, unbounded citizenship and the problematic nature of a transnational public sphere. *Political Communication*, 22(2), 179–196.

Carrillo, M. 2021. Affective Polarisation in Catalonia: A Case Study of the Electoral Campaign for the 2021 Regional Elections. Master Thesis. Master in Journalism and Innovation on Digital Contents. Universitat Autònoma de Barcelona.

Casero-Ripollés, A., R.A. Feenstra and S. Tormey. 2016. Old and new media logics in an electoral campaign. The case of podemos and the two-way street mediatization of politics. *The International Journal of Press/Politics*, 21(3). Doi: 10.1177/1940161216645340

Castells, M. 1997. La era de la información. *Economy, Society and Culture: The Power of Identity.* Vol. 2. Siglo XXI.

Castells, M. 2009. *Comunicación y poder.* Alliance.

Castells, M. 2012. *Networks of Indignation and Hope.* Alianza Editorial.

Celis, C. 2015. Towards an Immanent Critique of the Attention Economy. Labour, Time, and Power in Post-Fordist Capitalism. Doctoral Thesis. School of English, Communication & Philosophy. Cardiff University.

Celis, C. 2019. Notes on the political status of the image in the age of artificial vision. *Revista Barda*, 5(8), 89–106.

Coviello, L., Y. Sohn,, A.D.I. Kramer, C. Marlow, M. Franceschetti, N.A. Christakis and J.H. Fowler. 2014. Detecting Emotional Contagion in Massive Social Networks. *PLoS ONE*, 9(3), doi:10.1371/journal.pone.0090315.

Chomsky, N. 2011. Occupy the future. *In*: G. Muñoz Ramírez (Comp.). *Indignados.* Ediciones Bola de Cristal.

Eltantawy, N. and J.B. Wiest. 2011. The Arab Spring, Social Media in the Egyptian Revolution: Reconsidering Resource Mobilization Theory. *International Journal of Communication*, 5, 1207–1224.

Feenstra, R.A., S. Tormey, A. Casero-Ripollés and J. Keane. 2016. *The Reconfiguration of Democracy: The Spanish Political Laboratory.* Editorial Comares.

Fernández-Rovira, C. 2021. Political polarisation on the internet in times of pandemic: A case study of the Spanish tuitesphere between March and June 2020. Final grade thesis. Degree in Political Science and Administration. National University of Distance Education (UNED).

Fernández-Rovira, C. and S. Giraldo-Luque. 2021. *La felicidad privatizada. Information Monopolies, Social Control and Democratic Fiction in the 21st Century.* Editorial UOC.

Fuchs, C. 2014. *Social Media. A Critical Introduction.* Sage.

Fuchs, C. 2017. Dallas Smythe Today – The audience commodity, the digital labour debate, Marxist political economy and critical theory. Prolegomena to a digital labour theory of value. *In*: Fuchs, C. and Mosco, V. (Eds.), *Marx and the Political Economy of the Media* (pp: 522–599). Haymarket Books.

Fuchs, C. 2020. *Communication and Capitalism: A Critical Theory*. University of Westminster Press.

Giraldo-Luque, S. 2015. *Més enllà de Twitter. From indignant expression to political action*. Vic: Eumo.

Giraldo-Luque, S. 2018. Social protest and stages of moral development: An analytical proposal for the study of social mobilisation in the 21st century. *Palabra Clave*, 21(2), 469–498.

Giraldo-Luque, S., N. Fernández-García and J.C. Pérez-Arce. 2018. The thematic centrality of the #NiUnaMenos mobilization on Twitter. *Profesional De La Información*, 27(1), 96–105. https://doi.org/10.3145/epi.2018.ene.09. https://doi.org/10.3145/epi.2018.ene.09

Giraldo-Luque, S. and C. Fernández-Rovira. 2021. Economy of attention: Definition and challenges for the twenty-first century. *In*: Park, S.H., Gonzalez-Perez, M.A., Floriani, D.E. (Eds.), *The Palgrave Handbook of Corporate Sustainability in the Digital Era*. Palgrave Macmillan, pp. 283–305.

Habermas, J. 1981. *The Reconstruction of Historical Materialism*. Taurus.

Horta Ribeiro, M., R. Ottoni, R. West, V.A.F. Almeida and W. Meira. 2020. Auditing radicalization pathways on YouTube. *In*: *Proceedings of the 2020 Conference on Fairness, Accountability, and Transparency* (*FAT* '20*) (pp. 131-141). Association for Computing Machinery, Doi: https://doi.org/10.1145/3351095.3372879

Jones, J.J., R.M. Bond, E. Bakshy, D. Eckles and J. Fowler. 2017. Social influence and political mobilization: Further evidence from a randomized experiment in the 2012 U.S. presidential election. *PLoS ONE*, 12(4), e0173851. https://doi.org/10.1371/journal.pone.0173851

Katz, E. and P.F. Lazarsfeld. 1955. Images of the mass communications process. *In*: Katz, E. and Lazarsfeld, P.F. (Eds.), *Personal Influence: The Part Played by People in the Flow of Communication*. Free Press. pp. 15–42.

Kolhberg, L. 1992. *Psychology of Moral Development*. Desclée de Brouwer S.A.

Laraña, E. 1999. La construcción de los movimientos sociales. Alianza.

Lewis, P. 2017. 'Our minds can be hijacked': The tech insiders who fear a smartphone dystopia. *The Guardian*, October 6, 2017.

Lewis, R. 2018. Alternative Influence: Broadcasting the Reactionary Right on YouTube (White paper). Data & Society Research Institute.

Lewis, R. 2020. This is what the news won't show you: YouTube creators and the reactionary politics of micro-celebrity. *Television & New Media*, 21(2), 201–217. https://doi.org/10.1177/1527476419879919

Luhmann, N. 1995. *Power*. Anthropos.

Lynch, M., D. Freelon and S. Aday. 2017. Online clustering, fear and uncertainty in Egypt's transition. *Democratization*, 24(6), 1159–1177. http://dx.doi.org/10.1080/13510347.2017.1289179

Marcuse, H. 1987. *El hombre unidimensional*. Ariel.

McLuhan, M. 1969. *The Gutenberg Galaxy: Genesis of "Homo Typographicus"*. Aguilar.

Munn, L. 2019. Alt-right pipeline: Individual journeys to extremism online. *First Monday*, 24(6). https://doi.org/10.5210/fm.v24i6.10108

Ortiz, R. 2016. Social cybermovements: A review of the concept and theoretical framework. *Communication & Society*, 29(4), 165–183.

Ortiz, R. 2016a. Interpersonal and public communication strategies in social movements. Transformations of informal networks and repertoires in the age of the social web. OBETS. *Journal of Social Sciences*, 11(1), 211–254. Doi:10.14198/OBETS2016.11.1.09

Pariser, E. 2017. *The Filter Bubble: How the Web Decides What We Read and What We Think*. Taurus.

Piketty, T. 2014. *Capital in the Twenty-First Century*. Belknap Press.

Tarrow, S.G. 2016. *Power in Motion*. Alliance.

Torrego, A. and A. Gutiérrez. 2016. Seeing and tweeting: Young people's reactions to media representation of resistance. *Comunicar*, 47, 9–17. https://doi.org/10.3916/C47-2016-01

Touraine, A. 1969. *Sociology of Action*. Ariel.

Touraine, A. 2005. *A New Paradigm: To Understand Today's World*. Paidós.

Tufte, T. 2015. *Communication for Social Change*. Icaria.

Turkle, S. 2019. *In Defence of Conversation. The Power of Conversation in the Digital Age*. Attic of books.

Wiener, N. 1969. *Cibernética y Sociedad*. Buenos Aires: Suramericana.

Žižek, S. 2006. *Lacrimae rerum*. Debate.

CHAPTER

8

Reviving Topological Thinking in the Post-Media Condition

Justin Michael Battin

School of Communication & Design, RMIT University, 702 Nguyen Van Linh District 7, Ho Chi Minh City (Vietnam)

Introduction

The art critic Peter Weibel proposed that no single medium is dominant in the early twenty-first century, the different media available in the contemporary world mutually influence and complement each other. He refers to this phenomenon as the post-media condition, which is founded upon two phases: the equality of the media and the mixing of media. The first phase aimed to achieve equality between different forms of media, those considered traditional (like painting and sculpting) and those considered new and disruptive (such as photography, cinema, and digital art). As evidenced by the institutionalization of the latter forms, such as through university courses and the vast number of festivals held worldwide, that equality has been realized. The second phase, the mixing of media, involves how each medium and their affordances become not only connected to one another, but intertwined. Digital resources are utilized to augment the cinematic experience and museums encourage visitors to deploy their mobile devices to scan QR codes or enhance their understanding of the artistic work on display by way of augmented reality. Mobile media technologies enable people to contribute user-generated content to the ever-expanding media-scape (Appadurai 1996). The ultimate outcome of the post-media condition is the collective emancipation of the observer, visitor, and user; Weibel writes:

> *"In the post-media condition, we experience the equality of the lay public, of the amateur, the philistine, the slave and the subject. The very terms 'user*

Email: justin.battin@rmit.edu.vn

innovation' or 'consumer generated content' bear witness to the birth of a new kind of democratic art in which everyone can participate. The platform for this participation is the Internet, where everyone can post his or her texts, photos or videos. For the first time in history there is an 'institution', a 'space' and a 'place' where the lay public can offer their works to others with the aid of media art, without the guardians of the criteria" (Weibel 2012).

This vibrant display of optimism is not unfounded. Félix Guattari, for instance, in his initial conceptualization of what he saw as the decline of the mass media era, postulated that the transformation from a television dominated mass media society, one that produced a somnolent population[1], to a post-mass, hybrid-technology driven era will usher in an "era of collective-individual reappropriation and an interactive use of machines of information, communication, intelligence, art and culture" (Guattari 1996 [1990]). Enunciative assemblages, composed of tools made possible by infotech activity, in Guattari's view, would re-singularize passive and alienated audiences, thus enabling emancipation and autonomy. As stated by Genosko, in the post-media landscape, "the Guattarian subject is produced and punctuated by points of singularity and transformed by exploring the potential consistencies they bear in the process of inventing autonomy (2013).

In the twenty-first century, examples of this emancipation and autonomy are rife, from grassroots-organized protests to the amateur production and distribution of diverse multimedia across numerous channels. The most prominent arena where this emancipation and autonomy can most visibly be seen is on the internet, and especially the array of locative social media platforms available to frequent in the contemporary age, *places* of which are often praised for their seemingly egalitarian character, in the sense that only registration, a functional device, and internet service are the barriers to entry. However, as exposed by a variety of sources, from academic research (Figueiredo and Bolaño 2017) to popular documentaries (such as Netflix's *The Social Dilemma* [Orlowski 2020]), the algorithms facilitating interactivity with and on those platforms has invited intense skepticism. The optimism of the Web 2.0 era, which permitted user-generated content and invited participatory collaboration (see Jenkins 2006), has seemingly faded, as power has shifted from those who create and transmit information to

[1] This critique of mass media has roots in the Frankfurt School's notion of the capitalist driven culture industry. As stated by John Storey, "by supplying to means to the satisfaction of certain needs, capitalism is able to prevent the formation of more fundamental desires. The culture industry thus stunts the political imagination" (2013). Folk and grassroots cultures, those considered more authentic by way of practices beyond profit-driven infrastructures, are often subsumed into the culture industry. The commercialization process transforms this culture into another standardized purchasable good, thus depriving it of its critical function (Marcuse, 1968). In Adorno's critique, this results in passive consumption and eventually widespread conformity, as the produced materials shun differentiation and thus make no critical demands.

those who determine how information is stored, accessed, and distributed. Any egalitarian discourse about social media platforms and the technical devices used to access them are often met with an array of questions about privacy, censorship, manipulating information for the purposes of social control (Petrescu and Krishen 2020) and their role in perpetrating a post-truth world (Cosentino 2020).

These concerns, while significant impediments to the shared optimism conveyed by Weibel and Guattari, are indicative of a more widespread issue, the internet, and social media specifically, is a paradigmatic place perpetuating the collapse of meaningful distinctions. All information is flexible and all media objects are interchangeable. This chapter thus aims to demonstrate that while the promise of a post-media condition is not without hurdles, it remains pursuable through topological thinking, made possible by the articulation of place and place-making practices and a reinvigoration of a care-oriented perspective that centers the disclosive character of human beings. The first section of this chapter articulates the challenges facing the post-media condition through a Dreyfus-inspired reading of Heidegger, with the second devoted to addressing how topological thinking, as understood by Malpas (2006, 2017), and its association with unveiling one's potentiality for being, invites the possibility to reinvigorate a human being's disclosive agency and, subsequently, trust and confidence in the post-media condition.

Establishing Place through Spheres of Existence in the Post Media Condition

Drawing from Weibel's declaration, the internet, and locative social media specifically, is an institution, a space, and a place, through which individual people engage with a variety of media texts. The use of these specific words should not be overlooked. The first two, institution and space, have been recurring subject matter in the annals of social science research, and are typically positioned as structures or arenas which govern an individual's agency, and are therefore understood via power and power relations (see, for instance de Certeau 1984, Lefebvre 1974). Place, however, tends to be fathomed as "a meaningful location" (Cresswell 2004). Unlike the terms space and institution, place is grounded and substantiated through an experiential perspective (Tuan 1977). While there are numerous possibilities to characterize what constitutes a meaningful location, such as via physical location and locale, or the material setting for social relations, like Seamon's example of a *place ballet* (2015 [1979]), it is the notion of a sense of place, the embodied attachment people have developed for a place (Agnew 1987) that is most pertinent for consideration in the contemporary post-media condition. The reason for warranting priority to this aspect of meaningful location is first because of its immediacy, by way of how the notion conjures an understanding of *being-in*, and secondly because of how that immediacy, in the intimate sphere of the everyday, is seemingly threatened by a pervasive feeling of placelessness (Relph 1976). Locations imbued with meaning and belonging, those which Relph (1976) refers to as having *existential insideness*, have seemingly given way to a variety of *non-places* (Auge 1995); "places of transience rather than long-term attachment, places we pass through" (Malpas 2018 [1999]). Although Auge's critique is directed towards material structures shopping malls and airports, places representative of hegemonic globalization and time-space compression (Harvey 1989), it could equally apply to immaterial environments like social media platforms, given that each is assembled with a familiar structural layout and function, and are perpetually active and available for a quick (and equally superficial) traversal. The critique of electronic media through this place-oriented perspective is not necessarily new, as Meyrowitz (1985) has previously noted that the ubiquity of electronic media use stimulates widespread feelings of disembodiedness.

These critiques seem to implicitly reference concerns proposed by Martin Heidegger, who in "The Thing," suggested that contemporary technology's ability to abolish distances, materially and immaterially, brings no *nearness* (1971). With this statement, Heidegger is referring neither to calculable distance nor the subjective attachment one has for a place, as some of his critics contend,[2] but rather a nearness to *being itself* (Richardson, 2003). In this essay, Heidegger identifies a form of alienation both established and encouraged by a pervasive techno-logic, wherein all *things* are reduced to exploitable stock to ceaselessly manipulate for means-end purposes. A person's understanding of *being-in-the-world* is totally marginalized, as is their ability to acknowledge its manifestation.

Although Heidegger never seems to explicitly reference the notion of place, Malpas convincingly argues that the concept is a consistent thread throughout his philosophy; "Heidegger's thinking begins with the attempt to articulate the structure of a certain 'place.' The place at issue is not, however, any mere location in which entities are positioned, but rather the place in which we already find ourselves given over to the world and to our own existence within that world—the place that is one might say, the place of the happening of being" (2006). This statement clarifies why Heidegger is so critical for any analysis of place. Too often the notion of place is restricted to a brief moment in time-space; even Heidegger's Dasein, if interpreted quite literally as *being-there*, can imply place as static. A deeper reading of the philosopher's treatise, however, showcases a topological character, one which remains consistent throughout the entirety of his oeuvre (ibid). Through Heidegger, place becomes less associated with a particular spatio-temporal location, felt or otherwise, and instead as involving a range of interrelated constituent parts that collectively disclose (*Erschlossenheit*)

[2] Heidegger is too often painted as a Luddite by his critics, often due to his professed admiration and use of the German Romantic poets Friedrich Hölderlin and Rainer Maria Rilke and frequent references to his cottage in the Black Forest as an example of authentic dwelling. In "Building, Dwelling, Thinking," however, the philosopher is adamant that the references to the Black Forest farm house are not intended to imply that people should begin reconstructing such houses (1977 [1956]).

one's *being-in* (the world) and thus clears (*Lichtung*) a path for one's progression, illuminating within one's horizon their unique potentiality to be; as he writes: "only this clearing grants and guarantees to us humans a passage to those beings that we ourselves are not, and access to the being that we ourselves are" (1977 [1960]).

This conception of place begins in the philosopher's early work, as he argued that throughout the previous two centuries of occidental thought, a clear distinction between what constitutes subjects and objects has doggedly persisted. In *Being and Time* (1962 [1927]), Heidegger audaciously challenged this enduring presupposition of detached subjects standing over isolatable objects by inviting readers to reposition human beings as Dasein, or being-there. This reassessment of human beings emphasized the primacy of place by highlighting that, prior to ever assuming the perspective of a rational, calculating, or thinking entity, philosophical notions in vogue at the time, human beings were already fundamentally involved in a place imbued with meaning that is of much concern to the involved participant. Heidegger articulates this concern through Care (*Sorge*), and it is fundamental to his existential analytic of human being as Dasein, as it explains that unlike the detached Cartesian viewpoint dominant throughout modern western philosophy, which in itself is an heir to Plato's idealism, human beings are never indifferent to themselves and their world. For Heidegger, care "is to be taken as an ontological structural concept. It has nothing to do with 'tribulation,' 'melancholy,' or the 'cares of life,' though ontically, one can come across these in every Dasein. These, like their opposites, 'gaiety' and 'freedom from care', are ontically only possible because Dasein, when understood ontologically, is care" (Heidegger 1962 [1927]).

The care structure is united by three interrelated components, facticity, fallenness, and existentiality, and each is collectively needed to unveil being-in-place as meaningful. Briefly, facticity refers to the basic realities of existence, those taken-for-granted givens that cannot be altered. This notion of the care structure can be comprehended when a person astutely surveys their immediate surroundings and acknowledges they have been thrown into a world (*Geworfenheit*) imbued with circumstances and characteristics beyond their control. Although the observer may not have had any part in assembling that world, the objects populating it and the culture providing its fabric are accepted as being deeply important for that person's understanding of self. The second, fallenness, refers to a human being's tendency to embrace social norms established by *Das Man* (the anonymous one/they) and settle into a comforting conformity. The final, existentiality, refers to Dasein's own-most potentiality for being. To reference his use of the term *Ereignis* in *Contributions to Philosophy* (2012 [1989]), Heidegger posits that Dasein will experience an event, or moment of appropriation, wherein one's unique being *becomes an issue* (see Polt 2005). Dasein experiences a sort of *ekstasis* (*ἔκστᾰσῐς*), wherein a recognition of a being's manifestation, and consequently what one can possible be and what one can possible do, beyond the confines of the restrictive *Das Man*, becomes unveiled.[3] Through the care structure, Heidegger first demonstrates that human beings are grounded and delivered into a meaningful there, thus preceding any state wherein one fathoms themselves as a rational, calculating, or thinking being, as it is only through *that there*, whereby such fathoming could manifest. Second, by showcasing how the care structure illuminates the possibility for being, through existentiality, Heidegger demonstrates how Dasein is inherently topological, as it situates Dasein within a spatio-temporal moment with possibilities for further involvement and inhabitation cleared into view.

Among the most fruitful pathways to grasp the care structure and its role in illuminating place is through a phenomenological reading of the everyday interactions Dasein has with *things*, as these can revitalize an appreciation for the taken-for-granted background practices that indicate how one fluidly responds to the solicitations of the world, which are, according to Dreyfus (2015), fundamental for how one copes within an everyday context. In accordance with the argument initiated in *Being and Time*, things ought to be comprehended beyond observable objects at the behest of rational subjects. Rather, one must consider things ontologically, whereby they show themselves as referential to other things and collectively take part in constituting place, such as how, to use Heidegger's famous example, a hammer and nails referentially establish the setting for a workshop. A mobile media device provides an insightful example given our embodied relationship to them (see Ihde 1990; Battin 2017), their affordances of anyplace connectivity (Vanden Abeele et al. 2018), and the role they play in fostering and sustaining our media life (Deuze 2012). Whereas the majority of encountered objects have a uniquely specified role, mobile media devices are multi-purposed, and thus the principal interface for a range of interactions and engagements. Consider yourself for instance, by paying careful attention to the moment of solicitation, and how you are compelled to facilitate and nurture your engagement with that specific object in that specific moment, it is easier to comprehend how the happening of being is not only taking place, but also appropriated uniquely to you, to that which you are involved, and to that which you belong. As Dreyfus states, "Heidegger seeks to make us see that our practices are needed as the place where an understanding of being can establish itself" (2017). The rationale for this phenomenological approach, as stated by Mark Blitz, is because "we cannot construct meaningful distance and direction, or understand the opportunities for action, from science's neutral, mathematical understanding of space and time. Indeed, this detached and 'objective' scientific view of the world restricts our everyday understanding" (Blitz 2014).

In the work encompassing Heidegger's later career, he demonstrates a shift away from the "ahistorical, cross-cultural structures of everyday involved

[3] Of paramount importance, as well, is how the care structure reveals that human beings always find themselves invested in a social world, and that they are always socially constituted, although too often through Heidegger and his interpreters, this understanding is conceived in too monolithic a fashion (see Coeckelbergh,2018, p. 277).

experience" (Dreyfus 1991) and towards the urgent need to renew a receptiveness to being-in via illuminating our role in preserving the interconnectedness of its constituting elements. As referenced above, albeit briefly, he proposed that the contemporary age was deeply impoverished and a desolate time (Heidegger 2002 [1946]), as human beings have become engulfed in a paradigmatic *techno-logic*, alienated from being, and worse yet, unreceptive to its call. In accordance with the trajectory of detached (and disinterested) subjectivism of metaphysical thinking, of which Heidegger robustly critiqued first in *Being and Time*, human beings have entered into a historic stage characterized by nihilistic leveling, a notion he borrows from Nietzsche and Kierkegaard. Nothing remains from previous epochs, like deities or heroes/heroines, with any enduring authority. All things in the world, regardless of whether being material or immaterial, manifest as mere stock or standing-reserve (*Bestand*) in the sense that they lack any sense of sacredness; rather, they remain available to be called upon for purely commodifiable and exploitable use, and are thus easily replaceable. Moreover, meaningful distinctions, those moments, people, or *things* imbued with existential importance, fail to manifest as such and therefore fade into oblivion.[4]

Borgmann writes that this phenomenon can be observed during the transition between the modern to postmodern age. Drawing from Heidegger's critique of the technologic paradigm, he argues that postmodern technology (to which he appropriately refers to *soft technology*) not only draws things into the standing reserve, but eradicates the distinction of being objects entirely, as they are replaced by simulacra, endlessly transformable *information* under the control of arbitrary and flexible human desires (see Borgmann 1992). In Heidegger's own words, "cybernetics transforms language into an exchange of news. The arts become regulated-regulating instruments of information" (1972 [1969]). As noted by Dreyfus and Spinosa, Heidegger's understanding of postmodern technology moves beyond a subject controlling an object, such as what is currently observable in humanity's exploitation of nature. They posit, with reference to Nietzsche's influence, that Heidegger interprets a contemporary human being's will to power not as one's obtaining control over objects for the sake of satisfying subjective desires, but rather to produce and maintain flexible ordering; "thanks to Nietzsche, Heidegger could sense that, when everything becomes standing

---

4 Wrathall proposes a distinction between existential importance and instrumental importance. For Wrathall, things imbued with instrumental importance when they manifest as intelligible within a world structure. The inhabitant within that world understands how to wield it, as it manifests as ready-to-hand (*Zuhanden*) within a familiar context. Those things that manifest as existentially important are understood as "objects or person or practice … without which we would cease to be who we are. Such objects or persons or practices thus make a demand on us – require of us that we value them, respect them, respond to them no pain of losing ourselves" 2011). Wrathall further makes a connection to the Heidegger's rendition of nearness. Things with existential importance are critical conduits for Dasein's self-realization, as it is through such things that Dasein takes a stand.

reserve or resources, people and things will no longer be understood as having essences or identities" (Dreyfus and Spinosa 2017 [1997]). This idea manifests with the most clarity in his reading Nietzsche's eternal recurrence of the same, wherein a human being escapes from all communally informed values and norms to construct a life that is immensely rewarding and pleasurable, directed by one's subjective will (See Heidegger 1968 [1954], Heidegger 1982 [1961]).

Following Heidegger and Borgmann, Dreyfus and Spinosa suggest that the online environment is a definitive example of where the postmodern flexibility of information manifests. Using Kierkegaard's mid-1800s criticism of the public sphere, which was "characterized by a disinterested reflection and curiosity that level all differences of status and value" (2001), Dreyfus proclaims that the internet encourages people to become passive spectators to events, yet have an opinion about *everything*. Like the public sphere of Kierkegaard's age, people on the internet are prompted to share opinions regardless of its relevance to their community, their qualification, or worse, without having to demonstrate investment in or action on those opinions; "Kierkegaard saw that the public sphere was destined to become a detached world in which everyone had an opinion about and commented on all public matters without needing any first-hand experience, and without having or wanting any responsibility" (Dreyfus 2017 [2004]). This transformation, which disembodies users from a communal context, reduces the need for people to showcase any unconditional commitment towards that which grounds their identity. Rather, people perpetually engage in idle conversation (*Gerede*), a term Heidegger used to show that one understands things only approximately and superficially (Wrathall 2011). Whereas Weibel champions this newfound emancipation and democratization (to echo Habermas), Dreyfus proclaims that all meaningful differences distinguishing what is consequential for one's existence have been utterly levelled. One's Facebook, Twitter, or Instagram feed, for instance, may feature information related to the COVID-19 pandemic, entertainment media, promotional material for conservationist or human rights initiatives, content about a friend's recent vacation, or news about a locally held event. In Dreyfus' own words, "the highly significant and the absolutely trivial are laid together on the information highway" (2001).

Kierkegaard proposed this levelling would encourage people to plunge themselves into anything perceived as worth pursuing. He refers to this as the *aesthetic sphere* of existence, and in it, "people make enjoyment of all possibilities the center of their lives" (ibid). Deuze's aforementioned notion of a media life is perfectly suited to illustrate this phenomenon, given how users rapidly and indiscriminately surf through the online network consuming, commenting upon, and sharing a variety of different content with others. For Dreyfus, following Kierkegaard, this practice is anything but fulfilling, at least in any consequential sense, as the entire undertaking is founded upon diversion and evading one's deeper potentiality for being. In the aesthetic sphere of existence: "life consists of fighting off boredom by being a spectator at everything interesting in the universe and in communicating with everyone else so inclined. Such a life produces what we would now call a postmodern self, a self that has no defining content or continuity but is constantly taking on new roles" (ibid). Contemporary social media platforms provide examples of places where such role adoption transpires, as they both permit and encourage fluid identity performances. One must question, however, whether the word *performance* can even apply in this scenario. Social media activities best resemble curation; experiences, interests, and stories are meticulously selected to construct a continuously evolving narrative (see also Georgakopoulou 2017). An initial observation of these practices might raise no suspicion, as experimentation and negotiation are often considered crucial for the development of one's identity, particularly during the adolescent phase.[5] However, Dreyfus proposes that in the *aesthetic sphere* these performances and the stances expressed, labeled as virtual commitments, typically carry no risk in that they rarely impact the performer's status *offline*. One can profess their interests, share a meme, or make an inappropriate comment, and these would most likely have no bearing on their existence as a self; "the person in the aesthetic sphere keeps open all possibilities and has no fixed identity that could be threatened by disappointment, humiliation, or loss (2001).[6] Seemingly emancipated from the factical world, a person existing in the aesthetic sphere is liberated from making an unconditional commitment and thus risking losing who they are; if a person feels as though they have exhausted their interest in a given topic or would like to construct an alternate identity, then they are free to arbitrarily choose and eventually exhaust another. This cyclical dodging of ennui further demonstrates how possible interests (and even identities) have been relegated to mere flexible and transformable stock (*Bestand*). Moreover, as no interest (or identity) seems worth making unconditional commitments for, Dreyfus proposes that "individuals feel isolated and alienated. They feel that their lives have no meaning because the public world contains no guidelines" (2017 [1992]). Drawing

[5] Turkle initially praised the manner in which the internet allowed young people to articulate different identities (1995), but in recent literature has proclaimed that the conversations occurring in the digital realm ought to not be considered conversations at all, but rather connections (2011). She also notes that the inability to meaningfully converse has diminished levels of empathy and increased levels of loneliness (2016). One major factor to consider in her perspective is the adoption of the mobile phone, which has become the primary point of contact for many young people worldwide.

[6] This claim is certainly not without fault, given that social media users are situated worldwide, and often in countries with oppressive regimes utilizing social-media facilitated surveillance systems for policing purposes and social control. Although this seems most applicable in authoritarian nations, worldwide people have experienced cyberbullying over their shared content, which leads to the users experiencing *world collapse*, a term Heidegger uses to indicate an ontological death (which is different from physical demise). Aho explains that *world collapse* is a kind of "breakdown (*Zusammenbruch*) of meaning itself, where what dies or comes to an end is not a physiological entity but the ability to understand and make sense of the world and oneself." (2016).

on Bauman's critique of liquid modernity, Macey and McCauley (2021) echoing Dreyfus, observe that "we are defining ourselves according to our fandoms, our experiences, such as travel, our hobbies and our sexual preferences, not because they are more important than previous frameworks, but because they are all that is left to us" (2021).

This alienation eventually morphs into despair or *angst*, to use Kierkegaard's term, and prompts one to abandon the *aesthetic sphere* and transcend to the *ethical sphere*. To reference the Heideggerian care structure, this could be considered as acknowledging and acting on one's potentiality for being. As stated by Prince, in the ethical sphere "the self takes up a stable identity and an active stance on the issues or concerns that constitute his or her identity ... in this view, an online self would consistently make virtual commitments that resulted in involved actions and would therefore use social networks as tools to further his or her goals" (2018). The ethical sphere can thus be fathomed as a secular calling that seems difficult, if not impossible, to ignore, as it appeals to that which uniquely constitutes one's being-in-the-world. Online, organized fan activism is one case of virtual commitments melding with unconditional commitment in the ethical sphere. According to Jenkins, fan activism is a form "of civic engagement and political participation that emerge from within fan culture itself, often in response to the shared interests of fans, often conducted through the infrastructure of existing fan practices and relationships, and often framed through metaphors drawn from popular and participatory culture" (2012). Battin and Rystakova (2020) have showcased how members of the Harry Potter Alliance, a prominent fan activist group with chapters worldwide, are compelled to participate in the organization's various initiatives because of their commitment to social issues collectively acknowledged as urgent and the fan text through which the issues are comprehended. These activists appear to existentially recognize how taking a stance on *these issues* through *this text* permits them with a path *to be* in *this world*. Such a stance is significant because, as stated by Prince, "when we embrace our calling as unconditional commitment, our commitment discloses what will be the ultimate concern of our lives. Such unconditional therefore elude levelling and despair by determining what will show up as significant and insignificant, important and irrelevant, serious and lighthearted, on the basis of that which most matters in our life" (2018).

There are two points to further expound upon with respect to this phenomenon. The first of these is how existential projection into the ethical sphere manifests by way of engagement with *things*, and in particular through an ability to understand these things via their equipmentality, or the *referential nexus* that unifies them. In the case of the aforementioned example of the fan activists, specific interrelated things to consider are the different activists, or those who share a collective sense of urgency about the issues, the people who the movement aims to represent and address, the fan text itself, and also the media devices and platforms that are used as the loci to connect and communicate. The compulsion to participate and see the movement through to fruition is illustrative of what Malpas designates as the *happening of place*, a gathering of constitutive elements that allow for place to manifest as meaningful and worthwhile. In a chapter titled "The Poetry That Thinks" of his 2006 book, *Heidegger's Topology*, Malpas further articulates the experiential event (*Ereignis*) wherein our being is delivered over to us as a sort of happening, as a revealing that gathers us into a belongingness with being. He writes:

> *"The happening that is at issue here is not some abstract 'occurrence,' but a happening in which we are gathered into the concreteness and particularly of the world and to our own lives. As such, the happening at issue is also essentially a "there-ing," a "near-ing," a "place-ing"—it is a happening of that open region, that place, in which we find ourselves, along with other persons and things, and to which we already belong. In returning to the original Event that is the happening of belonging, the happening of being, we also return to the original happening of place."*

Malpas is keen to distinguish that this event is not intended to illuminate how one is located in a place through a purely spatial sense, a la the Cartesian tradition, but rather how it creates a founding wherein one becomes unified with the constituting elements of *their* world. Through this mutual appropriation, one can thus proceed *topologically*. For Dreyfus, this event of appropriation can also be further understood through the word *Augenblick* (translated often as *moment*, and usually with deeply religious connotations). The word is meant to indicate how one decisively responds to and within this founding, as one's potential *to be* depends upon their taking definitive singular action. With such decisive action, Dasein accepts responsibility *for their self*, and pursues its original role as a discloser, cultivator, and preserver of meaningful worlds (Spinosa et al. 1997).

Secondly, phenomenological manifestations of *being-in-the-world* in the current context must account for online environments and the technical devices[7] that permit access and sustain a person's telepresence therein. Firstly, human beings share an embodied relationship to these devices, as they are situated into what Kalaga (2010) refers to as the body's nebulous third; they are neither fully internalized nor externalized objects. This embodiment, in the sense that we wear them, carry them, and insert them into specific orifices, like our ears, has fostered an understanding wherein they lose the distinction of being foreign. Secondly, Luciano Floridi (2015) proposes that human beings have entered into a new

[7] Using the word devices is quite specific throughout this chapter, as it forges a connection with Borgmann's distinction between things and devices. With *things*, the world gathers and establish place, albeit in a way recognizable to the motivated inhabitant, thus preserving their involvement with being. Devices, conversely, are predicated on efficiency and thus conceal the happenings. In his own words, "what distinguishes a device is its sharp internal division into a machinery and a commodity produced by that machinery" (2009 [1984]).

revolution characterized by a blending of online and offline existence. He suggests that such distinctions are no longer possible, as mobile devices have blurred the ability to distinguish between real and virtual worlds and what constitutes human, machine, and nature. Furthermore, information is abundantly available and our constitution of being now is more predicated on a variety of interactions, processes and navigable involvement in and through networks. As stated by Verbeek (2015), Heidegger's notion of being-in-the-world ought to now be considered as *being-in-multiple-worlds*. Meyrowitz (2004) as well, seems to have revised his initial critical stance by suggesting that mobile media devices, rather than disembody people, give rise to *glocalities* (Meyrowitz 2004); while one may be situated in and engaged with an immediate locality, that person is simultaneously travelling virtually (Urry 2000). Each of these two points substantively impact one's topological involvement in the ethical sphere in the post-media condition.

As Kierkegaard notes, involvement in the ethical sphere on the internet can be undermined because one is always permitted to rescind their choice, and retraction often occurs particularly due to the allure of the anonymous *Das Man*. In the post-media condition, however, a new form of undermining has materialized. Drawing from Floridi's postulation that humankind has entered into the information revolution and are reappropriating themselves as connected, informational organisms, appropriated by a range of media technologies (2014), the ethical sphere is less undermined by choice and influence from social others, but *rather by the enframing paradigm of technology itself.* The data generated by our involvement with a range of media artefacts are profoundly shaping our worldview and the range of possible existential projections through which we can become involved (see Romele 2020). Floridi, for instance, proclaims that the information provided to the system is logged, analysed, and subsequently packaged and delivered in an accessible manner back to us while promising a more complete picture of *who we are*. While there are clear benefits to what this data can offer, such as its ability to identify information about bodily processes and more easily diagnose people to treat diseases, a point Floridi ardently stresses, we must be wary about how the data *generated* from our own behaviours are commodified for commercialization and control purposes (see Fuchs 2021 [2014]) and how we increasingly accumulate data for the sake of itself, a point often raised in discourses about our transformation into *quantified selves* (Hong 2020).[8]

Moreover, with the significant number of media tools logging our practices in the post-media condition, it is also worth questioning the role of *medium specificity*. This term, while typically associated with aesthetics and modernist and

8 As identified by Hong, the emergence of a quantified self is stripping away the liberal subject with embodied knowledge and sensibilities. Extracted data, rather than human judgement informed by experience, takes priority and is thus reshaping what qualifies as knowledge. Zuboff (2019) and Moore (2018) further argues that this data can be utilized for surveillance purposes in the workplace to maximize employee productivity.

postmodernist art theory (Chierico 2016), equally has a place in communication and media-oriented disciplines. McLuhan's phrase "the medium is the message," which prioritizes the communication medium over its delivered content as the catalyst for changes in social experience, is a widely accepted dogma. While content certainly plays a role, it is the medium (or *thing*), as an extension of human capacity (McLuhan 1964), that ultimately forms the horizon of human experience (see also Stiegler 1998). In the post-media condition, wherein technologies are wearable, portable, and perpetually logging our information, the significance of medium specificity is less clear. As mentioned in the opening section of this chapter, Guattari suggests that media, as tools, can be reappropriated by various subject groups to enable social practices for the re-articulation of the social (see Apprich 2013). Yet, as media platforms converge, it becomes increasingly difficult to distinguish one from another as it concerns form, purpose, and audience; in the post-media condition, media cease to be tools entirely, rather they have been systematically reduced to interconnected *hegemonic apparatuses,* wherein the flexible content delivered and communication forged is interpreted and valued based solely on its degree of measurability by shiftable and often arbitrarily chosen metrics like impact, reach, engagement, and so on. Given this collapse, which has been quite abrupt, one must question the loss of the ability to recognize meaningful distinctions not only between information, but also between different media.

The ultimate danger of this informational flexibility is how, when delivered back to users through a range of interconnected devices and utilized as the basis for how one should be, *the self thus ceases to be a self.* One's pursuit for singularity, the liberation from passivity and alienation, to use Guattari's terminology, is shaped by the endless treatment of the self as a customizable and optimizable resource informed by the aforementioned arbitrary and shiftable informational metrics. This is the ultimate example of the Nietzschean will to power and the total mobilization that Heidegger proposes is the *essence of technology*. He posits that, with the shift to computational technology, not only will objects disappear into the standing reserve, but human beings as well. Harkening back to the aesthetic sphere of existence, in an endless effort to optimize the self, humans freely (and often spontaneously) adjust *who they are* based on the continuous delivery of accumulated and catalogued data, which can range from engagement figures on social media to details about one's sleeping habits to or how many steps one takes per day. Again, while this data can be useful, as shown by Sharon and Zandbegen (2017), particularly to promote mindfulness, it equally promotes data fetishism, a belief that *data* is paramount for identifying truth and objectivity.

Thus, the issue standing before us is one of detached spectatorship wherein the quantified self becomes a project for maximum optimization and efficiency. Rather than approach human beings in its essence, a focus on the quantified self as a flexible project results in "unlocking, transforming, storing, distributing, and switching about ... but this revealing never comes to an end" (1977 [1954]). Instead of a self which takes an active stance on being that self within a world of revealed meaning unique to that Dasein, a la Kierkegaard's ethical sphere, the self in the post-media condition disappears within the paradigm of flexible information, all for the sake of greater efficiency and ordering. While Heidegger does clarify that the claim made by technology is still a mode of revealing *being-in*, human beings fail to recognize it as such. He refers this mode of revealing as *Das Gestell* (translated by Malpas as the framework) and considers it the ultimate danger, as it hides itself as a claim:

> *"The danger attests itself to us in two ways. As soon as what is unconcealed no longer concerns man even as object, but exclusively as standing reserve, and man in the midst of objectlessness is nothing but the orderer of the standing reserve, then he comes to the very brink of a precipitous fall, that is, he comes to the point where he himself will have to be taken as standing-reserve ... in this way the illusion comes to prevail that everything man encounters exists only insofar as it is his construct. This illusion gives rise to one final delusion: it seems as though man everywhere and always encounters only himself" (Heidegger 1977 [1954]).*

Within this framework, the things of the post-media condition world make no demands on us save for their arrangement within the paradigm of efficient ordering. One could consider this enframing as the triumph of Nietzschean nihilism, as humans are emancipated, albeit to identify and follow whichever standard manifests as valuable in a context of their own design. As it concerns the demise of the ethical sphere, as the claim of technology leaves people on unstable ground, they are thus unable to account for the topology of existence, and therefore with being itself. In the words of Kołakowski, "technology has gradually reduced our need for contact with being" (2008 [2004]). And yet, although this contact is seemingly inconsequential, place prevails, as Malpas identifies in a rather striking assertion. The world revealed, regardless of whether a technological paradigm or one considered by Heidegger as more authentic, still manifests "as constituting a certain *topos* and as accompanied by its own characteristic formation of places" (2006). In other words, the essence of technology is a mode of revealing establishing *place*, and is thus destined to undermine itself. As Malpas states, "Even in the face of technological ordering ... place endures ... so, in that endurance, does the possibility for another mode of revealing to come forth also" (ibid). Heidegger himself, equally recognizing this glimmer of resistance, remarked that "the closer we come to the danger, the more brightly do the ways into the saving power being to shine and the more questioning do we become. For questioning is the piety of thought" (Heidegger, 1977 [1954]).

Recovering Topology in the Post-Media Condition

As the essence of technology is in no way *technological*, Dreyfus suggests to appropriately account for the danger, one must first reflect upon the notion that "this threat is not a problem for which we must find a solution, but an *ontological condition* that requires a transformation in our understanding of being" (2017). Moreover, one cannot simply banish technology from their lives; Heidegger himself even acknowledges "it would be foolish to attack technology blindly ... we depend on technical devices; they even challenge us to ever greater advances" (1966). Instead, rather one must acknowledge that technical instruments, as things, are that which *constitute* human beings, as they contribute to the gathering of our belonging with beings, and thus that which makes the world and its meaning accessible to us. Kiran and Verbeek insightfully suggest that "in explicitly developing relations to technologies and their mediating roles, human beings give shape to their technologically mediated existence" (2010). They argue that a relationship with technology must be framed through a care-oriented perspective (particularly self-care), and that this framework offers a radical re-working of the Heidegger's notion of *Gelassenheit*, or releasement, in which contemporary technologies are permitted to enter our lives, and yet doubly remain outside; we use technologies, but disallow ourselves to succumb to their essence. The concept of *Gelassenheit*, developed late in the philosopher's career, is a sort of fundamental attunement (*Grundstimmung*) that permits, or rather invites, human beings to relate to other beings in their essence. The issue with this term, at least for me, is that it seems to minimize human agency, moreover, when shall one choose to leave them outside? This ambiguity reveals why Kiran and Verbeek's suggestion that explicit engagement with technology, albeit through a care-oriented perspective, is so critical, as it "entails both being involved with technology and taking a stance towards this involvement" (2010). They suggest that a care-oriented approach to engagement with technology permits us to first gain trust in the different technologies, as we are better able to distinguish and thus take account of their unique mediating roles, and secondly gain confidence, as "human beings are encouraged to take responsibility for the way in which one's existence is impacted by technology as well as to develop the skills to appropriate it in specific ways" (ibid).

In the post-media condition, this undertaking already appears to be well underway. The 2020 (and ongoing) COVID-19 pandemic has been pinpointed as a disruptor of routines and the inspiration for meaningful reflection; it has prompted people to acquire new skills and forge deeper (and more intimate) connections with others and themselves (Kapoor and Kaufman 2020). Furthermore, the stay at-home orders decreed by governments worldwide seem to have renewed interest in community, cultivating new skills and further developing one's creative faculties. From the perspective of care, there seems to be a turning away from challenging that which stands-forth, as enframing, and an appreciation for our role as cultivators, a rediscovery of existentially important objects and a renewed sense of place, particularly alongside others. The difficulty, however, will be sustaining this perspective. To do so, as with most Heideggerian analyses, one ought to be more attentive to language and its role in the manifestation and articulation of being-there.

In "The Question Concerning Technology," Heidegger posits to reflect upon technology's etymological origins, the Greek word *technē* (τέχνη), and proposes people must observe two principal notions with respect to the meaning of this word. First, for the last two thousand years, technology has been explicitly understood through only one of Aristotle's four causes (*causa efficiens*), which has thus inspired a restrictive narrowing of how technology is defined and thus understood (Heidegger, 1977 [1954]). In an attempt to reconcile the word with Aristotle's four causes (*causa efficiens, materialis, formalis*, and *finalis*), Heidegger recognizes that in its original context, "*technē* is the name not only for the activities and skills of the craftsman, but also for the arts of the mind and the fine arts. *Technē* belongs to bringing-forth, to *poiēsis* (*ποίησις*); it is something poietic" (1977 [1962]). With this statement, Heidegger demonstrates how the original meaning of the word accounts for how all four causes are wholly indebted to each other, and when they manifest to the involved participant, they do so together, intertwined and mutually constituted. In the philosopher's own words, "the four ways of being responsible *bring* something into appearance. They let it come *forth* into presencing" (ibid). The reconciliation of the four causes and returning of technology to the realm of *poiēsis*, or bringing-forth, rescues it from being purely defined through its instrumental and anthropological definitions, and renews it as a site for revealing and thus an opening for *topos*. This distinction of bringing-forth leads to the second point, being the link between *technē* and *epistēmē* (*ἐπιστήμη*). For the pre-Socratic philosopher, Xenophon, *technē* and *epistēmē* were indistinguishable terms, and "names for knowing in the widest sense" (ibid). Both were suitable vehicles for the unveiling of *alētheia* (*ἀλήθεια*), truth as unconcealment. The split between the two terms becomes most apparent in Aristotle's *Nicomachean Ethics*, as *technē* is distinguished from *epistēmē* "with respect to what and how they reveal" (ibid). The former is regarded as practical while the latter is positioned as theoretical. Heidegger, however, overwhelmingly rejects this distinction. To consider *technē* solely through a human act or as means undermines its true scope; "what is decisive about *technē* does not lie at all in making and manipulating nor in the using of means, but rather in the aforementioned revealing. It is as revealing, and not as manufacturing, that *technē* is a bringing-forth" (ibid). Heidegger's astute positioning of *technē* as *poiēsis* draws explicit attention to the constitution between technology (as a thing), the happening of place, and the ontological character of a human being.

Poiēsis is a term that appears infrequently throughout Heidegger's thought, but like Dasein's topological character, it presences without explicit reference. Although the word best translates as poetry today, in Ancient Greek it was the word used to describe the action of making. What is unique about this word above all the others that Heidegger uses to demonstrate the unveiling of being, is that through *poiēsis* one is able to better account for their specific role in the manifestation of the world. We can see how, as an integral part of the gathering, a transformation occurs wherein a new world is disclosed, nurtured by our own cultivating practices. In his late work, Heidegger suggested that it is through artists and poets that one can more explicitly recognize *poiēsis*, as they seem to be more open and attuned to worlds which compel them to not simply engage, but further sustain their involvement therein. As Inwood states:

> *"The artist or poet cannot do his work in any normal human way, in any way that already presupposes the world that he is to set up. He must be something like the vehicle of an impersonal force ... the artist must be "resolute", entschlossen, ecstatically "opened up" to this force (1997)."*

Such an *Augenblick*, however, need not be reserved for those categorized as artists, poets, or any other profession distinguished as a creative force. Dreyfus and Kelly, for instance, write that through *a meta-poietic* mindset, we transform into a craftsman of our own lives, although not to impose meaning on the world, per Nietzschean nihilism; indeed, "task of the craftsman is not to *generate* the meaning, but rather to *cultivate* in himself the skill for *discerning* the meanings that are *already there*" (Dreyfus and Kelly 2011). This *there*, of course, returns us to place.

An unorthodox, but nonetheless appropriate example of this cultivation of skillful discernment within a meaningful context can be seen in Dziga Vertov's *Man with a Movie Camera* (1929). Aptly regarded as a city symphony, the nearly one-hundred-year-old film presents a techno-utopia in the film's *diegesis*, whilst simultaneously challenging the traditional perspectives of the audience through the facilitation of the *Kino-Eye* (Michelson, 1984). Vertov appears to have two aims, the first being how the use of technological instruments enables new forms of looking (*through* cinema) and seeing (*through* montage editing). The film invites the audience to supplement their imperfect eye with a cinematic one, the aptly named *Kino-Eye*. With this embodied mergence, cinema is brought-forth not as an enframing device, but rather has a temporal happening that stimulates a new perspective and understanding, being an ethnographic-oriented unveiling of the literal assembling of a new society, facilitated by the collaborative use of technology for the purpose of forging communal unity (Tomas, 1992). Like other Soviet filmmakers of the era channelling the spirit of the age, Vertov encouraged a cinematic experience where the audiences were active participants rather than passive spectators. In accordance with this manifesto, the viewing audience becomes a partner, per se, in cinema's potential to temporally unveil new possibilities, to reconsider technology, to mold with it, for the manifesting of *new worlds*. Additionally, Vertov's choice to make himself and his camera a technologically facilitated hybrid-character in the diegesis, the audience is implicated to extend this understanding of technology and human beings as being conjoined into the wider social world, ultimately for the betterment of society. According to Tomas, "Vertov's unusual mode of collective observation and cinematic manufacture remain, to this day, one of the few coordinated attempts to design a 'social technology of observation' that could account for an expanding media culture while retaining and tactical, political, and social situational reflexivity" (Tomas, 1992). *Man With a Movie Camera* was produced during the Soviet Union's transition to a command economy, a period marked by rapid industrialization and collectivization, and the text demonstrates a deep understanding of the presence of technology during this time and the need to account for its potential *in establishing place*. Framed through Dreyfus and Kelly's remarks about craftsmanship, we can see how Vertov's Kino-Eye, according to Leyda, "was to make a new beginning" (1983) by breaking away from narrative trickery, to foster a cinematic movement of *film-truth* where plot was *detected* rather than *artificially invented* (see also Tretyakov 2006).

Nearly one hundred years post-Vertov, the post media condition's promise of Guattarian singularity is within reach, but it too must be uncovered rather than artificially constructed. In "The Origin of the Work of Art," Heidegger astutely notes that, prior to the institutionalization of art and its relegation to aesthetic enjoyment and a cultural activity, it illuminated the presence of deities and dialogues with human destiny. By returning to place as the site of being's opening, and recognizing our power to cultivate that illumination and chart a path to preserve it, *through the aid of contemporary technology as definitive in that constitution,* the singularity becomes one within our existentiality, guided by our care, with an eye towards meaningful action in the ethical sphere, bringing us ever closer to what Plato referred to as *ekphanestaton*, "that which shines forth most purely" (ibid). The question for our age and thus what defines the post-media condition, therefore, is one of topological questioning with careful consideration given to technology's role as a constituting agent; as Heidegger states, "the question concerning technology is the question concerning the constellation in which revealing and concealing, in which the coming to presence of truth comes to pass" (1977 [1960]).

References

Aho, K.A. 2016. Heidegger, ontological death, and the healing professions. *Medicine, Health Care, and Philosophy*, 19, 55–63. https://doi.org/10.1007/s11019-015-9639-4

Appadurai, A. 1996. *Modernity at Large: Cultural Dimensions of Globalization*. University of Minnesota Press.

Apprich, C. 2013. Remaking media practices: From tactical media to post-media. *In*: Apprich, C., Slater, J.B., Iles, A. and Schultz, O.L. (eds.), *Provocative Alloys: A Post-Media Anthology* (pp. 122–141). PML Books.

Battin, J.M. 2017. *Mobile media and poiēsis: Rediscovering How We Use Technology to Cultivate Meaning in a Nihilistic World.* Palgrave Macmillan. https://10.1007/978-3-319-59797-3

Battin, J.M. and E. Rystakova. 2020. Heidegger and fan activism: Unveiling the presence of poiēsis in contemporary online social mobilization. *Er(r)go: Theory-Literature-Culture*, 40(1), 63–83. https://doi.org/10.31261/errgo.7560

Blitz, M. 2014. Understanding Heidegger on Technology. *The New Atlantis: A Journal on Technology and Society*, 41, 63–80.

Borgmann, A. 1992. *Crossing the Postmodern Divide*. University of Chicago Press.
Borgmann, A. 2009 [1984]. *Technology and the Character of Contemporary Life: A Philosophical Inquiry*. University of Chicago Press.
Coeckelbergh, M. 2018. Skillful coping with and through technologies: Some challenges and avenues for a Dreyfus-inspired philosophy of technology. *AI & Society*, 34, 269–287. https://doi.org/10.1007/s00146-018-0810-3
Cosentino, G. 2020. *Social Media and the Post-Truth World*. Palgrave MacMillan. https://doi.org/10.1007/978-3-030-43005-4
Chierico, A. 2016. Medium Specificity in Post-Media Practice. *VIRUS 12*. http://www.nomads.usp.br/virus/virus12/?sec=4&item=6&lang=en
Cresswell, T. 2004. *Place: A Short Introduction*. Blackwell Publishing.
de Certeau, M. 1984. *The Practice of Everyday Life*. University of California Press.
Deuze, M. 2012. *Media Life*. Polity Press.
Dreyfus, H. 2001. *On the Internet*. Routledge.
Dreyfus, H. and S.D. Kelly. 2011. *All Things Shining: Reading the Western Classics to Find Meaning in a Secular Age*. Free Press.
Dreyfus, H. 2017 [1992]. Heidegger on the connection between nihilism, technology, art, and politics. *In*: Wrathall, M. (ed.), *Background Practices: Essays on the Understanding of Being* (pp. 173–197). Oxford University Press.
Dreyfus, H. and C. Spinoza. 2017 [1997]. Highway bridges and feasts: Heidegger and Borgmann on how to affirm technology. *In*: Wrathall, M. (ed.), *Background Practices: Essays on the Understanding of Being* (pp. 198–217). Oxford University Press.
Dreyfus, H. 2017 [2003]. Christianity without onto-theology: Kierkegaard's account of the self's movement from despair to bliss. *In*: Wrathall, M. (ed.), *Background Practices: Essays on the Understanding of Being* (pp. 231–246). Oxford University Press.
Figueiredo, C. and C. Bolaño. 2017. Social media and algorithms: Configurations of the lifeworld colonization by new media. *International Review of Information Ethics*, 26(12). https://informationethics.ca/index.php/irie/article/view/277
Floridi, L. 2014. *The Fourth Revolution: How the Infosphere is Reshaping Human Reality*. Oxford University Press.
Floridi, L. 2015. *The Onlife Manifesto: Being Human in a Hyperconnected Era*. Springer Open. https://10.1007/978-3-319-04093-6
Fuchs, C. 2021 [2014]. *Social Media: A Critical Introduction*. Sage Publications.
Genosko, G. 2013. The promise of post-media. *In*: Apprich, C., Slater, J.B., Iles, A. and Schultz, O.L. (eds.), *Provocative Alloys: A Post-Media Anthology* (pp. 14–25). PML Books.
Georgakopoulou, A. 2017. Sharing the moment as small stories: The interplay between practices and affordances in the social media-curation of lives. *Narrative Inquiry*, 27(2), 311–333.
Guattari, F. 2013. Towards a post-media era. *In*: Apprich, C., Slater, J.B., Iles, A. and Schultz, O.L. (eds.), *Provocative Alloys: A Post-Media Anthology* (pp. 26–27). PML Books.
Harvey, D. 1989. *The Condition of Postmodernity: An Enquiry into the Origins of Cultural Change*. Blackwell Publishing.
Heidegger, M. 1962 [1927]. *Being and Time*. Harper San Francisco.
Heidegger, M. 1966. *Discourse on Thinking*. Harper & Row Publishers.
Heidegger, M. 1968 [1954]. *What is Called Thinking?* Harper & Row Publishers.
Heidegger, M. 1971. The thing. *In*: Albert Hofstadter (ed.), *Poetry, Language, Thought* (pp. 164–180). Harper & Row Publishers.

Heidegger, M. 1977 [1960]. The origin of the work of art. *In*: Krell, D.F. (ed.), *Martin Heidegger: Basic Writings* (pp. 149–187). Harper San Francisco.
Heidegger, M. 1977 [1954]. Building dwelling thinking. *In*: Krell, D.F. (ed.), *Martin Heidegger: Basic Writings* (pp. 323–339). Harper San Francisco.
Heidegger, M. 1977 [1962]. The question concerning technology. *In*: Krell, D.F. (ed.), *Martin Heidegger: Basic Writings* (pp. 287–317). Harper San Francisco.
Heidegger, M. 1982 [1961]. *Nietzsche: Volumes Three and Four*. Harper Collins.
Heidegger, M. 2002 [1946]. Why poets? *In*: Young, J. and Haynes, K. (eds.), *Martin Heidegger: Off the Beaten Track*. Cambridge University Press.
Heidegger, M. 2012 [1989]. *Contributions to Philosophy (Of the Event)*. Indiana University Press.
Hong, S. 2020. *Technologies of Speculation: The Limits of Knowledge in a Data-Driven Society*. NYU Press.
Idhe, D. 1990. *Technology and the Lifeworld: From Garden to Earth*. Indiana University Press.
Inwood, M. 1997. *Heidegger: A Very Short Introduction*. Oxford University Press.
Jenkins, H. 2006. *Convergence Culture: Where Old and New Media Collide*. MIT Press.
Jenkins, H. 2012. 'Cultural Acupuncture': Fan Activism and the Harry Potter Alliance. *Transformative Works and Cultures*, 10(3). https://doi.org/10.3983/twc.2012.0305.
Kapoor, H. and J.C. Kaufman.2020. Making meaning through creativity during COVID-19. *Hypothesis and Theory*, 18. https://doi.org/10.3389/fpsyg.2020.595990
Kalaga, W. 2010. In/Exteriors: The third of the body. *In*: Front, S. and Nowak, K. (eds.), *Interiors: Interiority/Exteriority in Literary and Cultural Discourse* (pp. 3–23). Cambridge Scholars Publishing.
Kiran, A.H. and P.P. Verbeek. 2010. Trusting ourselves to technology. *Knowledge, Technology, & Policy*, 23, 409–427. https://doi.org/10.1007/s12130-010-9123-7
Kołakowski, L. 2008 [2004]. *Why is There Something Rather than Nothing: Questions from the Great Philosophers*. Penguin Books.
Lefebvre, H. 1974. *The Production of Space*. Blackwell Publishing.
Leyda, J. 1983. *Kino: A History of the Russian and Soviet Film*. Princeton University Press.
Macey, J. and B. McCauley. 2021. Mind games: Playtest as an allegory for liquid modernity. *In*: Duarte, G.A. and Battin, J.M. (eds.), *Reading Black Mirror: Insights into Technology and the Post-media Condition*. Transcript Verlag.
Malpas, J. 2006. *Heidegger's Topology*. MIT University Press.
Malpas, J. 2017 [2012]. *Heidegger and the Thinking of Place: Explorations in the Topology of Being*. MIT University Press.
Marcuse, H. 1968. *One-Dimensional Man*. Beacon Press.
Meyrowitz, J. 1985. *No Sense of Place: The Impact of Electronic Media on Social Behavior*. Oxford University Press.
Meyrowitz, J. 2004. New senses of place and identity in the global village. *In*: Nyiri, K. (ed.), *A Sense of Place: The Global and the Local in Mobile Communication* (pp. 21–30) Passagen.
Michelson, A. 1984. Introduction. *In*: Michelson. A. (ed.), *Kino-Eye: The Writings of Dziga Vertov* (pp. xv–lxi). University of California Press.
Moore, P. 2018. *The Quantified Self in Precarity*. Routledge.
Petrescu, M. and A. Krishen, A. 2020. The dilemma of social media algorithms and analytics. *Journal of Marketing Analytic*, 8, 187–188. https://doi.org/10.1057/s41270-020-00094-4

Polt, R. 2005. Ereignis. *In*: Wrathall and Dreyfus (eds.), *A Companion to Heidegger* (pp. 375–391). Blackwell Publishing.

Prince, C. 2018. Self-understanding in the age of the selfie: Kierkegaard, Dreyfus, and Heidegger on Social Networks. *In*: Battin, J.M. and Duarte, G.A. (eds.), *We Need to Talk About Heidegger: Essays Situating Martin Heidegger in Contemporary Media Studies* (pp. 117–150). Peter Lang Verlag.

Relph, E. 1976. *Place and Placelessness*. Pion.

Richardson, W. 2003. *Heidegger: Through Phenomenology to Thought*. Fordham University Press.

Romele, A. 2020. The dataficaton of the worldview. *AI & Society*. https://doi.org/10.1007/s00146-020-00989-x

Seamon, D. 2015 [1979]. *A Geography of the Lifeworld: Movement, Rest, and Encounter*. Routledge.

Sharon, T. and D. Zandbergen. 2017. From data fetishism to quantifying selves: Self-tracking practices and the other values of data. *New Media & Society*, 19(11), 1695–1709.

Spinosa, Charles, Flores, Fernando and Hubert L. Dreyfus. 1997. *Disclosing New Worlds: Entrepreneurship, Democratic Action, and the Cultivation of Solidarity*. MIT Press.

Tretyakov, S. 2006. Our Cinema. *October Magazine*, 118, 27–44. https://doi.org/10.1162/octo.2006.118.1.27

Tomas, D. 1992. Manufacturing vision: Kino-Eye, the man with a movie camera, and the perceptual reconstruction of social reality. *Visual Anthropology Review*, 8(2). https://doi.org/10.1525/var.1992.8.2.27

Tuan, Y. 1977. *Space and Place: The Perspective of Experience*. University of Minnesota Press.

Turkle, S. 1995. *Life on the Screen: Identity in the Age of the Internet*. Simon & Schuster.

Turkle, S. 2011. *Alone Together: Why We Expect More from Technology and Less from Each Other*. Basic Books.

Turkle, S. 2016. *Reclaiming Conversation: The Power of Talk in a Digital Age*. Penguin Books.

Urry, J. 2000. *Sociology Beyond Societies: Mobilities for the Twenty-first Century*. Routledge.

Vanden Abeele, M., R. De Wolf and R. Ling. 2018. Mobile media and social space: How anytime, anyplace connectivity structures everyday life. *Media and Communication*, 6(2), 5–14. https://10.17645/mac.v6i2.1399

Verbeek, P. 2015. Designing the public sphere: Information technologies and the politics of mediation. *In*: Luciano, F. (ed.), *The Onlife Manifesto: Being Human in a Hyperconnected Era* (pp. 217–228). Springer Open. https://10.1007/978-3-319-04093-6

Weibel, P. 2012. *The Post-Media Condition*. Metamute. https://www.metamute.org/editorial/lab/post-media-condition

Wrathall, M. 2011. *Heidegger and Unconcealment: Truth, Language, and History*. Cambridge University Press.

Zuboff, S. 2019. *The Age of Surveillance Capitalism: The Fight for a Human Future at the New Frontier of Power*. Profile Books Ltd.

CHAPTER
9

Ethical Insights for the Social Media Age

Cristina Fernández-Rovira

Universitat de Vic-Universitat Central de Catalunya, Carrer de la Sagrada Família, 7, 08500 Vic, Barcelona (Spain)

Introduction

For more than two decades, police departments in countries such as the United States, China, the United Kingdom, Germany, and Switzerland have been using predictive systems based on arrest data to determine crime hotspots in different cities. The main result of this, as pointed out by different studies (O'Neil 2017, Akpinar et al. 2021) is that the mathematical conversion of social activity into statistical data and correlations (Cardon 2018) generates geographical biases about specific places and demographics that is the poor, and slums. The consequence of the prediction about identified problem areas is that as surveillance and control over the population increases, so do arrests. A particular loop, a vicious circle within the application of artificial intelligence on the prediction of criminal behaviour and, as O'Neil (2017) highlights, a threat to equality and democracy.

According to Akpinar et al. (2021), who conducted a study on the application models of artificial intelligence in predicting criminal behaviour by neighbourhoods in large cities, even prediction systems based on reports (and not arrests) do not avoid social biases, mainly linked to poverty, in the identification of possible criminal behaviour. Saunders et al. (2016) explain that the firearm use and crime prediction systems used by the Chicago police in 2013 were completely unreliable. The algorithms employed did not help to lower crime rates and, as a negative consequence, black neighbourhoods were overrepresented in the identified crime lists.

Email: cristina.fernandez1@uvic.cat

In a similar study in 2019, Richardson et al. (2019), researchers at the AI Now Institute at New York University, pointed out that the algorithmic processing involved in police systems led to increased racial and social biases in the population. These researchers also drew attention to how the data was acquired, how it was handled and how some of the analyses were carried out to identify crime patterns. The lack of transparency of the data processing agencies that police forces used casts serious doubts about possible corrupt practices in the acquisition or capture of citizen data, as well as the legality of practices associated with the handling of the information: "Deploying predictive policing systems in jurisdictions with extensive histories of unlawful police practices presents elevated risks that dirty data will lead to flawed or unlawful predictions, which in turn risk perpetuating additional harm via feedback loops throughout the criminal justice system" (Richardson et al. 2019).

Nil-Jana Akpinar, a researcher in the Department of Statistics and Data Science at Carnegie Mellon University, stated in an interview with the Spanish newspaper El País (Pascual 2021) that her research applying the crime prediction models of companies contracted by the world's police led to very serious errors. Areas with high crime but low reporting, for example, were identified as low crime, while areas with high reporting but low crime were identified as predictably dangerous. Similarly, studies applying the predictive model in Bogotá, Colombia (one of the only cities with district-by-district data available), showed that some of the neighbourhoods needed to have half as many crimes as others for the system to identify them as high-crime areas. According to Akpinar, "data on crimes committed do not accurately reflect the actual distribution of crime and neighbourhoods have a different propensity to report to the police" (Pascual 2021), which leads to predictions being fed by spurious statistical relationships. For the Carnegie Mellon University researcher, the only way for predictive policing algorithms not to pigeonhole citizens in the most socially marginalised neighbourhoods "is not to use them" (Pascual 2021).

There are different companies that analyse the behaviour of citizens in their daily lives through the data they collect from different sources, and which converge on the internet, mainly Google and Facebook Inc (Olivo 2020). These companies, based on behavioural analysis, predict, for example, how a worker will perform, how much a user will be able to pay for health insurance, and how risky it is to grant health insurance to a person based on their health history or their Tik-Tok videos. One of the best-known cases of how data analytics can make important mistakes that are used for everyday decision-making is that of Catherine Taylor (O'Neil 2017, Pasquale 2016, Rubel and Clinton Castro 2021). ChoicePoint (now part of LexisNexis) conducted an analysis of Catherine Taylor's profile when she made a job application to the Red Cross. Taylor got a big surprise when the agency gave her a letter rejecting her application because ChoicePoint's data analysis for the Red Cross linked her name to the manufacture and sale of methamphetamines (Rubel and Clinton Castro 2021). Taylor had no criminal record, but her name was exactly the same as someone else's, from another city, who did have a criminal record.

Like the Red Cross, thousands of companies engage data analytics services for strategic decision making. Data analysts, such as ChoicePoint, report on automated processes that search for information in large, aggregated databases. There are no quality controls in these analyses (Rubel and Clinton Castro 2021). In fact, companies have so much information at their disposal that the use of error elimination filters, as in Taylor's case, would make it easy to identify (through their address or social security number, for example) that the information collected referred to two different people. Taylor discovered that as many as ten data analysis companies were generating erroneous reports on her background. It took her more than four years to find a job and she had trouble renting a flat (Pasquale 2016) even though the error was apparently corrected at the data analysis companies.

The ability of large technology companies to accumulate information, to sell it and to process it, defines the power of platforms to predict and inform future user behaviour. The prediction of behaviour (in purchasing, in the next click, in their political decisions, in their entertainment consumption) determines not only the economic power of large companies, but also the power of control over the individual and collective decision-making system. The above examples, related to the universe of crime prediction, as well as to the prediction of a worker's behaviour towards a job, demonstrates the high danger of the use of technology, mainly social media networks, as elements of data concentration that will be used both to influence the free will of citizens, manipulating their choices, and to determine the decisions that external institutions (such as the police or the Red Cross) can make about them.

As the examples have shown, the predictive power of algorithms is highly inaccurate and based on statistical operations that do not demonstrate proven causal relationships. At the same time, automated analyses have been shown to make serious errors and decisions based on them have an unfair impact on citizens (Akpinar et al. 2021, O'Neal 2017, Richardson et al. 2019).

The ethical universe that opens in the face of the direct impact on citizens, on their privacy, and on the alteration of their free will, and their freedom also raises different approaches to digital rights and to the regulatory frameworks that need to be introduced to rethink a new digital social contract.

This chapter takes an approach to how the practice of behaviour prediction (and the gradual loss of free will) has been introduced in large technological platforms, mainly in social media platforms. Likewise, it addresses the ethical perspective implied by the loss of individual and collective decision-making capacity (increasingly externalised in interfaces) and, finally, it takes a closer look at the new social contract based on a Charter of Digital Rights, a proposal that is beginning to be considered by different national and international public institutions, as well as by some civil society organisations.

The Path of Surrendering Free Will

One of the greatest elements of power of the big tech companies is their predictive ability. Predictive algorithms know everything about us because we have consciously or unconsciously given them our data, through contracts that we sign without reading. Our needs and tastes are not a problem for them, as we have established an emotional and familiar relationship with our devices and interfaces: A dependency. Our social media profiles are capable of projecting what we want or think we need at any given moment. This predictive calculation is key to generating profits for large technology companies such as Google, Facebook, or Amazon.

Although the power of technology to act as a prediction machine had been envisaged since the beginnings of computing linked to military conflicts (Turing 1974), it took some time for the development of computing to reach the great potential of reading and analysing information in real time. Moreover, it was necessary for the user to enter directly into the production of content and to provide, freely and automatically, all the information about himself or herself. The fall of the Berlin Wall and the introduction of the Internet in the 1990s (Fernández-Rovira and Giraldo-Luque 2021) made economic deregulation and access to information highways accessible to a large part of humanity. Only three decades after the launch of the World Wide Web, almost 60% of the population has access to the internet and 66% of them are smartphone users (We are Social and Hootsuite 2021). Every year, more than 300 million of the planet's inhabitants join the global connection.

The sum of users who, through millions of devices, began to create data on a massive scale gave rise to Big Data. This was joined by all the devices that automatically began to provide information about their users. Through the study of the Big Data generated on the internet, some of the data scientists, almost non-existent at the end of the millennium, began to carry out calculation and relationship exercises.

In fact, a researcher in artificial intelligence at the University of Washington, Greg Linden, wanted to work for a local company selling books on the Internet, back in 1997. Linden had already published some studies related to the prediction associated with users' trips based on travellers' reviews on the Internet. Linden called this first model *Automated Travel Assistant* (Linden et al.1997).

The owner of that bookstore was Jeff Bezos, who had a need: he wanted to recommend books to his customers. However, the data collected so far did not allow him to make recommendations that were truly relevant, as they focused on similarities that were not entirely accurate. The customer could come to the same conclusion on his own because no value was added to the purchase decision (Mayer-Schönberger and Cukier 2013).

Then, Linden entered the scene and proposed to the bookstore, known as Amazon, to find associations between products sold through collaborative, item-by-item filtering (Linden et al. 2003). That patent revolutionised predictions: "Many applications use only the items that customers purchase and explicitly rate to represent their interests, but they can also use other attributes, including items viewed, demographic data, subject interests, and favourite artists. At Amazon.com, we use recommendation algorithms to personalize the online store for each customer. The store radically changes based on customer interests, showing programming titles to a software engineer and baby toys to a new mother" (Linden et al. 2003).

Of course, the algorithm was not only for book recommendations, but could be used for any product. Amazon saw the opportunity to become a shop that could offer every consumer good, imaginable in the market. The company also fired the workers who wrote the reviews of the books it was trying to sell, because the predictive power, which was based on customer data, was so great that it exceeded all expectations. Amazon had found a way to work with all the available data and thus predict a person's specific taste. Today, every major technological company uses the prediction system, which accounts for up to three quarters of the companies' revenues (Giraldo-Luque and Fernández-Rovira 2021).

Amazon's predictive capability became a competitive advantage because small or even large companies did not have the capacity at the time to use big data as a predictive factor. As Mayer-Schönberger and Cukier (2013) point out, many websites can now recommend products, content, friends and groups, even if they do not know the reason why people might be interested in them. Obviously, it would be interesting to know that reason, but it is of no importance when it comes to growing sales. "Knowing what, however, drives clicks" (Mayer-Schönberger and Cukier 2013).

In the early 2000s, Facebook and Google refined their platforms and services to maximise the potential of Linden's model. From 2007 onwards, as the financial crisis based on industrial models deepened, Google took over all user browsing (with its Android operating system and Chrome browser) and Facebook Inc. understood that the value of the emotionality of friendships and, later, likes and followers, could determine future user behaviour.

Today, a large part of humanity (53% of the population uses social media networks, according to the 2021 report by We are Social and Hootsuite) has developed an emotional attachment to technologies, especially the smartphone. The portable device that is inseparable for humans, with an average use of more than 6 hours a day among young people (Giraldo-Luque and Fernández-Rovira 2020), it concentrates a good part of the power of a few companies, of the technological giants. Amazon, Google, the social media platforms, and another small number of companies concentrated in Silicon Valley (United States) and Shenzhen (China) control and manage the data of most of the world's population. With this data, and in real time, they develop a high probability of getting predictions right and of proposing reliable predictions to external companies. A study conducted by a team of IBM researchers showed that the traces people leave on social media can delineate their psychological profile to predict their behaviour. The study shows how the analysis of tweets and messages on Facebook can define a person's personality (Zhou et al. 2013).

Other studies have indicated that more than 61.5% of users are willing to share their personality traits on social media and that, from them, there is a very high feasibility to automatically infer a user's personal behavioural traits from their behaviour on social networks (Gou et al. 2014).

To the prediction system we must add the Page-Rank algorithm developed by Google. This system, which establishes the level of popularity of web pages according to the citations they receive, allowed Google, as a monopoly company, to control the advertising and knowledge markets. The company, which has already shifted towards personalised search results made possible by coercing the user to create a personal Google account (without which it is almost impossible to use a digital device). Page Rank uses the information to establish the hierarchy of results on the basis of data from millions of users. By shifting its system towards predictions of user consumption, Google ended up assigning a specific value, for each user, to each of the sites that make up the web. In Pasquinelli's words: "Google established its own hierarchy of value for each internet node and thus became the first systematic global rentier of community intellect" (2009).

The Dominance of Freedom of Decision as an Ethical Exercise: A Systematic and Subtle Exercise of Power

Through predictive systems that guide citizens' decisions, the user-consumer reduces his/her uncertainty and feels confident about his/her choices. In the end, it can be seen as a prediction made based on their own personality (or at least the one reflected in social media). But what kind of decision is this? Is it a free or a conditioned decision? Of course, the debate of people's free will goes beyond what has been called algorithmic culture, but if one assumes that the platforms know everything, then is there no one more equipped than social media to make our own decisions? The affirmative answer to the question would mean that we are entering a kind of oligopolistic and controlled society, where a few companies control all available information, but also a society where contradictions disappear, because are assimilated and controlled by predictions.

Something that also disappears in the *Amazon model* is the explanation of the causes of phenomena. The analysis of massive data finds correlations (Cardon 2018), but no explanations of what affects us as individuals and as a society. Moreover, as the algorithm decides for us, it also prevents deep thinking from taking place. We abandon ourselves to the easy decision, the one that comes predetermined, pre-designed according to our behaviour on the screen. We don't even find the point of asking ourselves the *why* of things. The focus of interest is the *what*, because it is the basis of the predictive system: What do you want, what do you like, what do you desire, what do you need? But it is never asked why we want, why we like, why we desire and why we need what appears on the screens. In fact, when asked qualitatively about why, the answers induce irrational behaviour: “sometimes you feel the need to be there, to go on Instagram and look at the pictures, even though you’ve seen them hundreds of times. It’s a need”, stated a student interviewed in the study developed by Giraldo-Luque and Fernández-Rovira (2020). Statistical, algorithmic correlations have ended up eliminating the complexity of the reality that surrounds us.

However, the phenomenon of simplification and the reduction of complex thinking, as well as of available options, does not seem to worry us too much. In general, individuals feel reassured because the control structure that forms the oligopoly of technological platforms can assimilate opposing interests. This control structure feeds on our behavioural data, on the very thought expressed, and creates mechanisms to integrate difference, so that there are no oppositions that are dangerous for the system that the platforms themselves represent: it becomes “a hyper-technological feudalism in a democracy without citizens” (Burgaya 2021).

Prediction systems standardise thoughts, feelings, and aspirations. Through them we live satisfied and preconditioned. We do not even find refuge in culture, since the *platformised* cultural model also responds to the society of control. Human contradictions are not staged in art, theatre, or music because predictive mechanisms anticipate uncertainty and offer consumer goods that relax the tensions once represented in cultural creations. Culture is presented as domesticated by the filters of control and censorship imposed, in a non-transparent and random way, by digital platforms.

The pioneering companies were able to establish an information oligopoly by accumulating predictive power from the collection and management of user data. Thus, a few companies can now offer a wide range of predictions. Furthermore, their capacity to collect such data is growing, so that they learn more and more about the user-consumer behaviour. As with the example of crime prediction data used by different police forces around the world, the data feedback loop is used to refine the mechanisms for projecting and satisfying user needs. It will become easier and faster to predispose and anticipate their purchase, their vote, their pre-made decision.

Predictive power has become one of the great challenges in today’s society, where the control of the meaning of life is in question. Philosophers Michel Foucault (1979) and Niklas Luhmann (1995) approached the study of power relations understood as the horizon of individual and collective actions from a functional (Foucault) and structural (Luhmann) perspective. The power of prediction concentrated in the dominant platforms of the internet today renews the approaches to power of both authors, as we are witnessing the creation of a subtle monopoly, or oligopoly, that exercises a universal dominion over the decision-making capacity of citizens, and which is invisible, as Foucault announced in his description of power (Fernández-Rovira and Giraldo-Luque 2021).

Power exercised invisibly has also been internalised by users. We have become accustomed to it and today it is assumed as something vital. This was stated by some of the young people interviewed in previous research:

"Sometimes you don't even know what you're looking at, but you feel the need to be there, on the screen, just scrolling".

"They have created a certain fear in us. We feel lonely if we are outside the networks, and even more so if we are without a mobile phone. I think everyone is afraid of being nobody".

"The other day I had to leave my mobile phone at home because it was out of battery and when I went out, I was afraid..., something might happen to me" (Giraldo-Luque and Fernández-Rovira, 2020).

Invisible to the user, the process in which consumers send data on their digital behaviour at zero cost and every time they use their smartphone has been automated. In this way, control over prediction, control over visibility prices and control over the reception, channelling and administration of data has been consolidated, which exemplifies the model of social media platforms (Giraldo-Luque and Fernández-Rovira 2021). All the data collected also serve to further adjust the algorithms that determine the consumer's interests (Rahwan 2018) and feed the control exercised over the user at every second (Žižek 2018).

As Turkle (2019) and Fuchs (2012) argue, on the internet the user is always under surveillance, even if he or she has apparently freely accepted the rules of use of each application. However, large companies exert ideological control because there is no alternative outside of them. According to Fuchs (2017), platforms execute their power through the emotional coercion they exert on people to use their products. Thus, it is increasingly difficult not to be connected, as social pressure points in that direction. Moreover, established power relations are perpetuated by algorithmic prediction-driven selections, which make users dependent.

The above scenario entails ethical and citizens' rights issues. Indeed, technological advances have historically been accompanied by debates around their effects on people, their possibly addictive designs, and their social impact. For example, regarding the main debates on artificial intelligence, Müller (2021) points to privacy and surveillance; behavioural manipulation; opacity of systems; bias in decision-making systems; human-robot interaction; automation and employment; autonomous systems; machine ethics; artificial moral agents; and singularity or superintelligence. The changes now brought about by new technologies are such that critical voices are beginning to emerge calling for greater respect for sensitive issues (Fernández-Rovira et al. 2021).

Some former employees of leading tech companies have spoken out about the technologies they helped design. Chamath Palihapitiya, former vice president of user growth at Facebook, commented in 2017 at a conference at Stanford Business School that she felt "tremendous guilt" for having developed "tools that helped tear at the social fabrics through which society functioned" (Wong 2017). The former Facebook worker also said that "there is no civil discourse, there is no social cooperation, there is only misinformation and lies" and, perhaps most alarmingly, the former Facebook worker warned the listening public that their behaviours are being "programmed" even if they don't realise it (in Wong 2017). For his part, Sean Parker, one of Facebook's former presidents, denounced in 2017 that the company was working directly on people's psychological vulnerability: "It's a social-validation feedback loop... exactly the kind of thing that a hacker like myself would come up with, because you're exploiting a vulnerability in human psychology" and that "it literally changes your relationship with society, with each other. It probably interferes with productivity in weird ways. Only God knows what it's doing to our children's brains" (Solon 2017).

One of the creators of virtual reality and former advisor to Google and Microsoft, Jaron Lanier, who now advocates that we should all delete our social media profiles, compared Facebook to a priest who confesses two billion people and then sells their data to the highest bidder (in Fortson 2018). Lanier coined the term *Bummer machine* to refer to a system that involves the domination of user behaviour, which serves to build an empire that is then sold or rented out (Fernández-Rovira and Giraldo-Luque 2021).

Additionally, former Google strategist James Williams has publicly stated that the tech industry is the "largest, most standardised and most centralised form of attentional control in human history" (in Lewis, 2017). Having created the metrics for the development of Google's global search advertising markets, Williams is now critical of the fact that they can manipulate the decisions and actions of internet users. "I realised: this is literally a million people that we've sort of nudged or persuaded to do this thing that they weren't going to otherwise do", Williams stated in 2017 (Lewis 2017).

Lights Going Out: The End of the Free Will

Williams' concern is not science fiction. The Cambridge Analytica scandal showed that the use of user information on social media can be used to manipulate or directly influence a decision, in this case, an electoral one. It can be used to predict. But this example was not the only one (Giraldo-Luque 2015). Bond et al. (2012) demonstrated the influence of messages received by Facebook users on subsequent decision-making. In the experiment, Facebook intentionally manipulated the messages that 60 million users received on their walls. The results showed a direct impact on 60,000 voters and an indirect impact on 280,000 citizens. The experiment was replicated in a study published in 2017 which concluded that: "these results replicate earlier work and they add to growing evidence that online social networks can be instrumental for spreading offline behaviours" (Jones et al. 2017).

In a similar study, Coviello et al. (2014) demonstrated the incidence of emotional contagion under the influence of messages received by users on their social networks. It is a conclusion also demonstrated by Kramer et al. (2014) who noted, after an analysis of almost 700,000 Facebook users, that "emotional states can be transferred to others via emotional contagion, leading people to experience the same emotions without their awareness. We provide experimental evidence that emotional contagion occurs without direct interaction between people".

Studies have shown the power of social media to use the data collected from users to influence their decision-making and even their state of mind. This power of incidence, the power of orientation (Luhmann 1995), confers the ability to predict behaviour based on the information inputs received. The concern about predictive capacity becomes a warning when it directly interferes with the orientation of the will of free individuals and, therefore, with the normal functioning of democracy away from the passions of the masses (Williams 2021).

The power of social media technology acts as a catalyst for human attention. The focus on attention is relevant because, according to Williams (2012), it is attention that empowers individuals to "manage the will at all levels of human experience". Williams draws on John Stuart Mill (2016) who argues that human freedom depends initially on the internal mastery of consciousness, as well as the freedom to think and feel. This freedom, for Mill, also requires the conscious possibility of choice and definition of tastes and inclinations, the "freedom to arrange our lives according to our mode of being" (Mill, 2016, in Williams 2021). According to Williams (2021), "it is significant that the first of these freedoms is that of the mind, on which freedom of speech depends. And freedom of speech is meaningless without freedom of attention, which is both its complement and prerequisite".

To govern one's will, or freedom, makes it possible to direct one's individual and collective life. It is the condition for determining what one wants to become, the goals one plans to achieve and the path one follows to reach them. On the social or political level, the construction of will, which manifests in the form of collective choice, also determines authority and democratic legitimacy. In the normative terms of the Universal Declaration of Human Rights, democracy is exercised through the will of the people, as the basis of the authority of public power (Williams 2021). Thus, the control of will, not exercised freely or constructed through the manipulated orientation of individuals' choices in a sophisticated and mechanised way, is also an attack on democracy.

According to Williams, accepting the importance of will in structuring political life implies admitting that the platforms that concentrate the attention of users (Fernández-Rovira and Giraldo-Luque 2021) have a broader objective than the economic one: "the attention economy is a project that ultimately aims to shape the very basis of our politics. The product is not only the user but also the political person, the citizen" (Williams 2021).

Following James Williams' proposal, the loss of autonomous control of citizens' will manifests itself through three metaphors, each of which refers to a verb of the human constitution as an individual and collective project. Firstly, the *focal light*, associated with the verb to do, constitutes the "immediate faculties with which we orient our consciousness and our activity towards a given task. It enables us to do what we want to do" (2021). As an attack on focal light, social media networks bombard everyday life through notifications (Giraldo-Luque et al. 2020), a system that frames the concept of "functional distraction" (Williams 2021).

By diverting the individual from their initial frame of action and perceptual information, the constant and automated functional distractions associated with emotional behaviour (Giraldo-Luque et al., 2020) also distract individual awareness away from reflective information. Through persuasive design, sensory inputs of attention-grabbing appeals monopolise users' actions so that they focus on dominant digital spaces. By eliminating the possibility of reflexive information, critical capacity is also lost, and thus the will loses its first assault to surrender to serialised behaviour that responds to a call for attention, a notification. In the domain of attention, "the dimming of the focal light can frustrate political participation in several ways. Firstly, because it makes us refractory to political information and prompts us to read information of another kind (...). On the other hand, the kind of distraction that prevents us from being politically informed may have been deliberately contrived by the propaganda of a political party or some other interested sector (...). Such strategic distraction manoeuvres can also be employed to divert the focus of political debate" (Williams, 2021), as we currently see in almost 100% of the world's political communication systems.

Secondly, *the astral light*, associated with the verb to be, guarantees the "general faculties to navigate our lives guided (...) by certain values and life goals. It enables us to be who we want to be" (Williams 2021). The construction of identity, determined by the constitutive values of the personality, as well as by the effect of being able to tell a life story associated with those life paths, has been fragmented into 9-second episodes (Patino 2019) that, necessarily, must be appreciated by other users to exist. The narrative meaning of life and identity is, nowadays, the impact that a publication achieves on social media platforms. By limiting the value of identity and narrative construction to the dynamics of the networks, the unique (and disruptive) identity is lost and ends up being homogenised socially, culturally and politically. "When the astral light dims, we find it harder to be who we want to be. We feel the self-fragment and disintegrate, plunging us into a more existential kind of distraction" (Williams 2021).

The universe of behavioural conditioning of the platforms ends up guiding the individual towards areas that do not really correspond to his or her personality, but which, when socially assumed, imposed by the platform, become routines or socially accepted and defended values. Behaviours often do not represent the values and principles of the user-citizens, but they themselves declare that they are not willing to stop using the platform that conditions their actions (Giraldo-Luque and Fernández Rovira 2020). What is most worrying is that the weakening of the astral light leads to the valuing, legitimising, and extolling of trifles (Williams 2021), which ends up hooking billions of people. In fact, the predominant values in the behavioural construction of social networks have been associated with narcissism (Gnambs and Appel 2018) and personal fame (Uhls and Greenfield 2011), while promoting aesthetic operations as a cause of social pressure (Hughes

2017), the need to receive likes and emotional rewards (Fernández-Rovira and Giraldo-Luque 2021).

Williams warns of the consequences of the gradual diminution of the astral light of individuals as a collective. For the author, in the face of the loss of individual identity, it is also possible to project the forgetting of shared identity, that is, to be less clear about what "unites us to the rest of society" (Williams, 2021). In fact, Williams defines it as a deeply irrational disunity (beyond polarisation) that is characterised as the loss of identity (Giraldo-Luque 2018), as the discordance of the deep self among the political class, and as the birth of a society whose only habit (in the absence of a collective project) is indecision (James 1989). Along with Rousseau (2011), Williams takes up the warning that the seduction of particular interests (today represented in a few dominant technological companies) can easily turn collective decisions away from the general will. This is a very powerful and subtle control that "is most often exercised through a subdivision of society into groups that end up renouncing their membership of the larger group. In extreme cases, the divergences between different groups are accentuated to the point where their isolation is self-reinforcing. When this identity division becomes moral, it leads to a deeper tribal delegitimization that leads to a particular kind of populism" (Williams, 2021).

However, prediction avoids the feeling of frustration that indecision can generate, as the decision is already programmed for the individual and is conveniently delivered via a screen notification. The user happily accepts it, while retweeting it and, with it, further destroying his or her autonomous ability to decide and rescue his or her identity.

Finally, *daylight*, related to the verb to know, forms "the fundamental faculties (...) that enable us to determine values and life goals. They enable us to want what we want to want" (Williams 2021). Defining an objective and a goal in terms of identity and value construction framed everything that characterized society until we entered the 21st century. Not knowing how to define an objective about what we want to be in an autonomous way is an epistemic problem and therefore affects the capacity to articulate, defend and channel a system of beliefs or opinions that make up a body of knowledge and a basic scenario of rational knowledge and justifications. The loss of reflection, of (now externalised) memory capacity, of autonomous prediction and planning, of calmness, of logic (Williams 2021) and of the search for intersubjective argumentation as the principle of consensus (Habermas 2005) has ended up undermining the most important aspects of the social and democratic edifice.

The imposition of principles and values (linked to the neoliberal and individualistic practice that is also promulgated by the leaders of large technological companies) through platforms takes precedence over traditional principles and values linked to the social, the discursive and the reflective. In fact, the epistemic problem of attention (Williams 2021) hinders the integrated association of experiences of very different orders to detect the common structures that are shared (Miller and Buschman 2014). On the contrary, it fragments the social and even identity to exercise a more effective control framework in its individuality. By setting aside the common structures that "constitute the abstractions, general principles, concepts and symbologies that are the substratum from which the sophisticated and holistic thinking necessary to establish authentic long-term goals is born" (Miller and Buschman 2014). It is their instincts and automatisms induced through emotive notifications that determine the new common sense. Williams cites Harry Frankfurt (2006) who likens this shift to "unrestraint", or the moment when deliberate or reflective reasons justifying actions are suppressed and, as is the case today, the way is opened for the rule of impulsivity and emotionality.

The description of the epistemic problem is described by Williams (2021) in terms of two processes. Firstly, the author indicates that the control/prediction framework of social media distracts the individual from knowledge of himself/herself and his/her external world. It encloses them in small layers of banal information specially designed to attract the user's attention. Without this information, without this knowledge, the individual loses the meaning of life, and the objective is traced by a statistical convergence of data. "I'm there because everything is easier, because that's where everyone is", users often say. Secondly, Williams (2021) proposes that epistemic distraction, which is connatural to the dynamics of the social media platform, affects autonomy and dignity: "It violates the integrity of the self, subverting the conditions necessary for it to exist and thrive, leaving it without a ground on which to stand". In this sense, studies on the affections generated by social media networks report that their use is associated with social anxiety, depression, or discouragement (Giraldo-Luque et al. 2020, Lin et al. 2016).

The diminished capacity for reflection ultimately determines the impossibility of questioning and restructuring habits, beliefs, and thoughts. Applied to the exercise of citizenship, such behaviour, fragmented and manipulated according to notifications of interruption and de-concentration, has a major impact on people's political behaviour. Determining "what we want to want" is no longer an autonomous or deliberately discursive construction. On the contrary, it becomes a value acquired through sounding boards (Pariser 2017) and states of happiness that limit the capacity for social change. For Williams, it all comes down to the emotive and explosive frame. From venerating indignation (Castells 2012), we have taken an important qualitative leap to see it as a danger in the face of the impulsive and unethical preponderance of moral indignation (Williams 2021) as an everyday practice of social media.

The social network mechanism (Giraldo-Luque et al. 2021) prescribes a basically sterile outrage behaviour as it does not generate a collective thought or a stable discursive identity. According to Williams, the easily controllable social network outrage gives rise to chain reactions of such a dimension that the transition from outrage to reflection is impossible. "The result is ochlocracy, the rule of the plebs, of the law of the street (...). The process of ochlocratic justice, fuelled by moral outrage and viral chain reactions, is a capricious, arbitrary and uncertain process" (Williams 2021). The process is the imaginary or placebo effect of the law of the street. The government is exercised by the platform, which acts as a tyranny legitimised by the users, who have ceded their free will.

Towards a Digital Bill of Rights: The 21st Century Social Contract

Faced with the challenges of the digital society, some countries have put in place measures to guarantee citizens' rights in the digital sphere. Chile is one of the pioneering countries in the world. Currently, the South American country is about to include neuro-rights in its Constitution, with the aim of protecting mental identity from manipulation. For its part, the current Spanish government has proposed collective work on a Charter of Digital Rights, which includes a catalogue of rights that should be protected in the digital environment, such as the right to data protection, identity in the digital environment, the right not to be traced and profiled, and the right to digital inheritance, among other issues.

Although not legally binding, in 2019, Canada, Denmark, Estonia, Israel, South Korea, Mexico, New Zealand, Portugal, the United Kingdom and Uruguay signed the Digital Nations Charter, committing to work towards open-source systems, high quality connectivity and supporting citizens in the use of digital services, among other things.

The European Union is also working on the Digital Services Act, which aims to protect consumers and their fundamental rights online, establish a system of accountability for online platforms and boost innovation, growth, and competitiveness in the Single Market (European Commission 2020). On the other hand, in 2018 the European Union adopted the General Data Protection Regulation, which preserves citizens' personal data and allows the free flow of data and guarantees the right to erasure on the internet. In the United States, on the other hand, there is no federal data protection law and the right to be (online) forgotten is a matter of debate (Moreno Bobadilla 2019).

From Columbia University in the United States, Professor Rafael Yuste, who is also spokesperson for the Morningside Group, leads a team of academics who advocate the need to protect this type of rights. In the same vein, the Israeli thinker Yuval Noah Harari, in a conference at the Mobile World Congress held in Barcelona in 2021, warned of the "extremely dangerous" situation in our society, in reference to the fact that the world is turning into colonies of data, raw data, flowing towards the "imperial centres in China and the United States, where algorithms, artificial intelligence and the most sophisticated tools of control are developed" (Del Castillo 2021). According to Harari, society may drift towards a "digital authoritarianism" that could be avoided if the data collected on people were used to help them, not to manipulate them; if data were not allowed to be concentrated in only one or two places in the world; and if as surveillance of individuals increases, surveillance of people by governments and corporations increases proportionally.

At the same time, civil society organisations have begun to rally to defend their own rights. In one of the most recent examples, the Campaign for a Commercial-Free Childhood (now called Fairplay) spoke out against Facebook's initiative to create an Instagram aimed at children under 13. In a letter sent to Mark Zuckerberg signed by more than 30 organisations listing more than 25 scientific studies warning of the dangers of social media for children, adolescents and young people, the organisation pointed out that: "A growing body of research demonstrates that excessive use of digital devices and social media is harmful to adolescents. Instagram, in particular, exploits young people's fear of missing out and desire for peer approval to encourage children and teens to constantly check their devices and share photos with their followers. The platform's relentless focus on appearance, self-presentation, and branding presents challenges to adolescents' privacy and wellbeing" (CCFC 2021).

Likewise, in 2017, the organisation NOYB-European Center for Digital Rights succeeded in ensuring that Facebook had to keep its users' data on European territory in order to guarantee that the management of users' information complied with the European community's privacy standards. NOYB has also submitted a notification in May 2021 for more than 500 European companies to adapt to the EU's General Data Protection Regulation. The organisation, led by lawyer and activist Max Schrems, has demanded that companies modify the use of digital *cookies* that remain on devices every time we access a page. Once on devices, *cookies* collect information about user navigation and, after being analysed by digital engines, are used to personalise digital advertising, while they can also be sold to third party companies to feed more databases for commercial and predictive purposes.

The above examples, both from the public sphere and from organised sectors of civil society, show that fighting against the unethical dominance of large platforms is not impossible. The Charter of Digital Rights in Spain, as well as European initiatives or those of other countries in the world, point out some pillars that need to be highlighted and instilled in citizens who use the internet and in particular social media daily. Citizens must ensure that their rights are respected, at least based on the following principles (Burgaya 2021) associated, in many cases, with the fundamental Charter of Human Rights:

- *Recognition of property rights on the Web*: It is essential to establish a clear notion of ownership over private data and user-generated content (which cannot be exploited without express consent and without receiving compensation in return). At the same time, it is essential to ensure user ownership of the tangible and intangible goods (hardware and software) they purchase, and to establish the regulatory impossibility of access to and control of user data through purchased services. Property rights are the basis of the capitalist system, and, without them, it would also not be possible to guarantee ownership of data for the companies that exploit it commercially.

- *Recognition of intellectual property rights in platforms for publishing user-generated content*: The principle of exploitation of copyright based on the creation of content (intellectual, artistic, journalistic, informative, etc.) must prevail over any appropriation and exploitation of content by technological platforms. All content created must recognise the authorship of its users and, in the event of exploitation, the authors must always receive financial remuneration previously agreed with the company exploiting the content.
- *Recognition of the right not to leave a footprint on digital platforms*: Digital footprints form the material through which predictions about user behaviour are established. Users should be able to decide at any time during their use of digital platforms, in which spaces and for what periods of time their data and behaviour and personal information collected during their browsing can be recorded and used. Any automatic data collection procedures by websites, such as *cookies*, should be absolutely forbidden.
- *Recognition of the right to erasure of existing information about the user on digital platforms:* Users should be able to access, through a simple procedure (such as when registering on a social network) and transparently, all available history about him or her that is stored by technology companies and, at the same time, have the power to request the erasure of all or part of the information that the platform holds and uses for any purpose.
- *The recognition of the right not to be monitored by any computer and/or technological system:* Unless explicitly and fully informed consent is given by users, internet platforms and hardware and software devices should not be able to record, listen, watch, track, trace, monitor or otherwise record users' activities, content, movements or data capture, especially when these are done without the user using the applications or devices or knowingly connecting to the internet. This right should apply to any type of digital tool, device, or application.
- *Recognition of the right to privacy*: Despite Mark Zuckerberg's belief that "privacy is no longer a relevant social norm" (Turkle 2019), the right to privacy is one of the guarantors of democratic relations and is also included in many national constitutions. In the Spanish case, for example, Article 18 of the Constitution states that: "1. The right to honour, to personal and family privacy and to one's own image is guaranteed. 2. The home is inviolable. No entry or search may be made therein without the consent of the owner or a court order, except in the case of flagrant offence. 3. The secrecy of communications and, in particular, of postal, telegraphic and telephonic communications is guaranteed, except in the case of a court order. 4. The law shall limit the use of information technology to guarantee the honour and personal and family privacy of citizens and the full exercise of their rights" (Article 18. Spanish Constitution).
- *Recognition of the right to honour*: Honour, which is also recognised in Article 12 of the Universal Declaration of Human Rights, can also be violated on the internet and, above all, on social networks. Harassment practices and mistakes

made in the administration of data for decision-making by external agents that affect individual users determine a direct violation of citizens' rights. Both users and the platforms through which such activities are channelled also deserve to be regulated for the benefit of people's good reputations. As Josep Burgaya (2021) points out: "the law should act to prevent theft, insult, plagiarism, slander and falsehood from having free reign".

- *Recognising children's rights and strengthening child protection*: Children are exposed daily to inappropriate content and digital practices that threaten their physical, identity, psychological and moral integrity. Pornography, violence, access to and publication of photographs and personal data, and practices such as harassment, bullying or sexting deserve special attention to protect minors. In everyday life, there are areas of restricted access to minors; these are conventions acquired for their protection. Social networks cannot be the exception and, without any doubt, the proposal of Instagram for minors, as well as the access of minors to social media platforms using false identities, deserve to be socially and judicially sanctioned.
- *Recognition of labour rights*: It is now widely accepted that digital workers never disconnect from their work. They are available 24 hours a day, seven days a week and have to answer messages and answer their mobile devices constantly in their break times. It is important to rethink workers' rights that have been lost with the introduction of digitalisation of companies and that prevent, for example, family conciliation or stability in working hours (O'Neil 2017). At the same time, protecting the privacy of workers and job applicants is crucial. Their measurement and characterisation by artificial intelligence and big data systems should be prohibited or at least seriously regulated.
- *The regulation and limitation of advertising*: While advertisements, their content and space are effectively regulated on television and radio, the internet and social networks are rife with the abuse of messages and advertisements that rely on the user's private behaviour. "It is imperative to control invasive marketing systems, the use of illegally obtained personal data, spamming methods, misleading advertising, the blurred world of advertising compensation... The use of personalised profiles based on our data should, as a matter of freedom, simply be prohibited" (Burgaya 2021).
- *The promotion of the right to disconnect*: Just as there are smoke-free zones, spaces where access to polluting cars has been banned, or places where alcohol consumption is limited, there should also be internet-free zones. While (economic) progress was measured a few years ago through network access, social progress can start to be measured through the restructuring of the social fabric through technology-free conversation and interaction (Turkle 2019, Burgaya 2021).
- *The adaptation of electoral laws to the new digital challenges*: The case of Cambridge Analytica, the proliferation of fake news, the massive disinformation systems that feed on Twitter, Facebook and Tik-Tok and the

latest studies that relate political and affective polarisation to the use of social networks (Fernández-Rovira 2021) merit a normative reflection on the use of social networks in electoral campaigns. "The protection of citizens from the immense possibilities of polarisation (...). The use of personal data, campaigns instrumentalised from digital platforms, the phenomenon of post-truth, the role of remote-controlled bots (...), if all this is not stopped, electoral contests will only be a spectacle and democracy will be turned into an absolute farce" (Burgaya 2021). In predictive terms, digital profiling has a power of precision that is very important for political campaigns. With a few likes on social platforms like Facebook, biometrics and psychometrics can determine what activates like-minded voters with precision. It can also determine what kind of actions can turn off adversarial voters. In the end, democracy, the democracy we live in today, has declined on the real-time measurement and prediction of the impact a message can have on social media. Its influence has come to determine the state of opinion of the citizenry, positioned in irreconcilable camps. Therefore it is necessary to adapt the regulations related to the electoral process to oblige platforms and their users to respect the principles that govern democracy, especially during election periods.

Conclusion: Giving in on the Trifle to Return to Freedom

The ethical discussion centred on the predictive power of social networks points directly to the battle for the recovery of citizens' free will. To the liberation of their autonomous will. To the rescue of their freedom. As technological platforms have invaded and hijacked people's everyday relationships and actions, their ability to think, reflect and decide for themselves has been transferred to the realm of algorithms. It is they, the automatic calculations, that end up deciding what an individual should do in each instant (after each notification), but also what projection of the future they should have (in terms of the construction of values), how they should think and in which groups they should participate.

The prediction generated by data capture of free expression, of free movement, of the privacy of each user, directly alters the concepts of humanity (as an autonomous perception of the construction of will) and of citizenship (as a collective projection of a horizon of a communal future). The short-circuit established by social media platforms to personal and collective reflection, to the capacity for deliberation, introspection and reasoning, has weakened to the maximum the common identity affinity of the public. In the social network, the paradox is that the self is king.

The predictive mechanisms that delineate the individual will no longer seek only the economic control they already have. The struggle of totalitarian entrepreneurs, like the conquest and appropriation (for sale) of space, is centred on the domination of the will, of agency. It is striking that Jeff Bezos, the bookseller, said after his adventure in space, in July 2021, that he thanked all those who had paid for his trip. Nobody started a revolution (not even a sign of moral indignation on Twitter) at his words.

Social networks have made their prediction lines necessary for users. They have made them dependent and as different studies have shown, that in the absence of connection, end of battery life or damage to the smartphone, panic in the face of uncertainty comes into play. *Off-line* users are completely disoriented, lost and nervous. Humanity has been annihilated and the user has been turned into a fickle, fragile, manipulable, and fickle non-subject, without will and without objectives (either individual or collective), without a future. He/she is an isolated individual.

The challenge posed by technology in the dimension of the social restructuring of the individual and the citizen, in the direct affectation of democracy, is centred on a new social contract based on a bill of digital rights. The recovery of traditional rights, recognised in most of the world's constitutions, which have been violated by large technology companies that also violate the tax and labour laws of many countries around the world, must be recognised as a collective identity, as a social objective. But the first problem is that building such a collective identity will be impossible if society's behaviour is controlled and predictive, as it is now.

The second problem is perhaps more complex as it concerns the lost will of the citizen. Nobody is willing to give up the triviality, the (supposed) privileges and facilities granted by social media networks and big technology companies in exchange for gaining rights over privacy, honour, and freedom of expression, good information or the right balance in a market economy free of monopolies. Giving up these benefits in exchange for rights seems impossible in the information age, just when the most information is available and accessible. In the 21st century there are very few people willing to give up the empire, they are already happy to have their own processed, sold, and analysed data make decisions for them.

References

Akpinar, N.J., M. De-Arteaga and A. Chouldechova. 2021. The effect of differential victim crime reporting on predictive policing systems. *In*: *Conference on Fairness, Accountability, and Transparency* (FAccT '21), March 3–10, 2021, Virtual Event, Canada. ACM, New York, NY, USA, 17 pages. https://doi.org/10.1145/3442188.3445877

Bond, R.M., C.J. Fariss, J.J. Jones, A.D.I. Kramer, C. Marlow, J.F. Settle and J.H. Fowler. 2012. A 61–million-person experiment in social influence and political mobilization. Nature, 489, 295–298. doi: https://doi.org/10.1038/nature11421

Burgaya, J. 2021. *La manada digital. Feudalismo hipertecnológico en una democracia sin ciudadanos*. pp. 302-304, El viejo topo.

Cardon, D. 2018. *Con qué sueñan los algoritmos: Nuestras vidas en el tiempo de los Big Data.* Dado Ediciones.

Castells, M. 2012. *Redes de indignación y esperanza.* Alianza Editorial.

Coviello, L., Y. Sohn, A.D.I. Kramer, C. Marlow, M. Franceschetti, N.A. Christakis and

J.H. Fowler. 2014. Detecting emotional contagion in massive social networks. *PLoS ONE*, 9(3), https://doi.org/10.1371/journal.pone.0090315

Campaign for a Commercial-Free Childhood (CCFC) 2021. *Tell Mark Zuckerberg: No Instagram for kids!* https://fairplayforkids.org/wp-content/uploads/2021/04/instagram_letter.pdf

del Castillo, C. 2021. Los tres consejos de Harari para frenar el "autoritarismo digital". *Eldiario.es*. June 28, 2021. https://www.eldiario.es/tecnologia/tres-consejos-harari-frenar-autoritarismo-digital_1_8083352.html

European Commission. 2020. Proposal for a regulation of the European Parliament and of the Council on a Single Market for Digital Services (Digital Services Act) and amending Directive 2000/31/EC. Brussels, December 15, 2020. COM(2020) 825 final. 2020/0361(COD). https://eur-lex.europa.eu/legal-content/EN/TXT/HTML/?uri=CELEX:52020PC0825&from=es

Fernández-Rovira, C., J. Álvarez Valdés, G. Molleví and R. Nicolas-Sans. 2021. The digital transformation of business. Towards the datafication of the relationship with customers. *Technological Forecasting and Social Change*, 162, 120339. https://doi.org/10.1016/j.techfore.2020.120339

Fernández-Rovira, C. 2021. La polarización política en internet en tiempos de pandemia. Estudio de caso de la tuitesfera española entre marzo y junio de 2020. Final grade thesis. Grado en Ciencias Políticas y de la Administración. Universidad Nacional de Educación a Distancia (UNED).

Fernández-Rovira, C. and S. Giraldo-Luque. 2021. *La felicidad privatizada. Monopolios de la información, control social y ficción democrática en el siglo XXI*. Editorial UOC.

Fortson, D. 2018. Interview: Jaron Lanier's 10 reasons why you should delete your social media accounts right now. *The Sunday Times*. May 20, 2018. https://www.thetimes.co.uk/article/interviewjaron-laniers-10-reasons-why-you-should-delete-your-social-mediaaccounts-right-now-6qwzpg2px

Foucault, M. 1979. *Microfísica del poder*. Ediciones de la Piqueta.

Frankfurt, H. 2006. *La importancia de lo que nos preocupa: Ensayos filosóficos*. Katz Editores.

Fuchs, C. 2012. Google Capitalism. *Triple C. Communication, Capitalism & Critique*, 10(1), 42–48. https://doi.org/10.31269/triplec.v10i1.304

Fuchs, C. 2017. Dallas Smythe Today – The audience commodity, the digital labour debate, marxist political economy and critical theory. Prolegomena to a digital labour theory of value. *In*: Fuchs, C. and Mosco, V. (Eds.), *Marx and the Political Economy of the Media* (pp. 522–599). Haymarket Books.

Giraldo-Luque, S. 2015. *Més enllà de Twitter. De l'expressió indignada a l'acció política*. Eumo Editorial.

Giraldo-Luque, S. 2018. Protesta social y estadios del desarrollo moral: Una propuesta analítica para el estudio de la movilización social del siglo XXI. *Palabra Clave*, 21(2), 269–498. https://doi.org/10.5294/pacla.2018.21.2.9

Giraldo-Luque, S. and C. Fernández-Rovira. 2020. Redes sociales y consumo digital en jóvenes universitarios: Economía de la atención y oligopolios de la comunicación en el siglo XXI. *Profesional de la información*, 29, 5, e290528. https://doi.org/10.3145/epi.2020.sep.28

Giraldo-Luque, S., P. Aldana-Afanador and C. Fernández-Rovira. 2020. The struggle for human attention: Between the abuse of social media and digital wellbeing. *Healthcare*, 8(4), 497. https://doi.org/10.3390/healthcare8040497

Giraldo-Luque, S. and C. Fernández-Rovira. 2021. Economy of attention: Definition and challenges for the twenty-first century. *In*: Park, S.H., Gonzalez-Perez, M.A., Floriani, D.E (eds.), *The Palgrave Handbook of Corporate Sustainability in the Digital Era* (pp. 283–305). Palgrave Macmillan.

Gnambs, T. and M. Appel. 2018. Narcissism and social networking behavior: A meta-analysis. *Journal of Personality*, 86(2). 200–212. https://doi.org/10.1111/jopy.12305

Gou, L., M.X. Zhou and H. Yang. 2014. KnowMe and ShareMe: Understanding automatically discovered personality traits from social media and user sharing preferences. *In*: *Proceedings of the SIGCHI Conference on Human Factors in Computing Systems (CHI '14)*. Association for Computing Machinery, pp. 955–964. https://doi.org/10.1145/2556288.2557398

Habermas, J. 2005. Tres modelos de democracia: sobre el concepto de democracia política deliberativa. *Polis: Revista de la Universidad Bolivariana*, 4(10).

Hughes, D. 2017. Social media pressure is linked to cosmetic procedure boom. *BBC News*, June 22, 2017. https://www.bbc.com/news/health-40358138

James, W. 1989. *Principios de psicología*. Fondo de Cultura Económico.

Jones, J.J., R.M. Bond, E. Bakshy, D. Eckles and J. Fowler. 2017. Social influence and political mobilization: Further evidence from a randomized experiment in the 2012 U.S. presidential election. *PLoS ONE*, 12(4), e0173851. https://doi.org/10.1371/journal.pone.0173851

Lewis, P. 2017. Our minds can be hijacked: The tech insiders who fear a smartphone dystopia. *The Guardian*, October 6, 2017. https://www.theguardian.com/technology/2017/oct/05/smartphone-addiction-silicon-valley-dystopia

Lin, L.Y., J.E. Sidani, A. Shensa, A. Radovic, E. Miller, J.B. Colditz, B.L. Hoffman, L.M. Giles and B.A. Primack. 2016. Association between social media use and depression among U.S. young adults. *Depression and Anxiety*, 33(4), 323–331. https://doi.org/10.1002/da.22466

Linden, G., S. Hanks and N. Lesh. 1997. Interactive assessment of user preference models: The automated travel assistant. *In*: Jameson, A., Paris, C. and Tasso, C. (eds.), *User Modeling: Proceedings of the 6th International Conference, UM97*, Springer, Wien, Vienna, New York.

Linden, G., B. Smith and J. York. 2003. Amazon.com recommendations: Item-to-item collaborative filtering. *IEEE Internet Computing*, 7(1), 76–80. https://doi.org/10.1109/MIC.2003.1167344

Luhmann, N. (1995). *Poder.* Anthropos.

Mayer-Schönberger, V. and V. Cukier. 2013. *Big data. La revolución de los datos masivos*. Turner.

Mill, J.S. 2016. *De la libertad.* Acantilado.

Miller, E.K. and T.J. Buschman. 2014. Neural mechanisms for the executive control of attention. *In*: Nobre, A.C. and Kastner, S. (eds.), *The Oxford Handbook of Attention.* Oxford University Press.

Moreno Bobadilla, A. 2019. El derecho al olvido digital: Una brecha entre Europa y Estados Unidos. *Revista de Comunicación*, 18(1), 259–276. https://dx.doi.org/10.26441/RC18.1-2019-A13

Müller, V.C. 2021. Ethics of artificial intelligence and robotics. *The Stanford Encyclopedia of Philosophy* (Summer 2021 Edition). Zalta, E.N. (ed.). https://plato.stanford.edu/archives/sum2021/entries/ethics-ai/

Olivo, L. 2020. FAANGS Out: What big tech wants with your data. *humanID.* https://human-id.org/blog/faangs-out-what-big-tech-wants-with-your-data/

O'Neil, C. 2017. *Armas de destrucción matemática. Cómo el Big Data aumenta la desigualdad y amenaza la democracia*. Capitan Swing Libros.

Pariser, E. 2017. *El filtro burbuja: Cómo la red decide lo que leemos y lo que pensamos*. Taurus.

Pascual, M.G. 2021. Algoritmos de predicción policial: Para qué se usan y por qué se ensañan con los más pobres. *El País*, July 21, 2021. https://elpais.com/tecnologia/2021-07-21/algoritmos-de-prediccion-policial-para-que-se-usan-y-por-que-se-ensanan-con-los-mas-pobres.html

Pasquale, F. 2016. *The Black Box Society: The Secret Algorithms That Control Money and Information*. Harvard University Press.

Pasquinelli, M. 2009. Google's page rank algorithm: A diagram of cognitive capitalism and the rentier of the common intellect. *In*: Becker, K. and Stalder, F. (Eds.), *DeepSearch: The Politics of Search Beyond Google* (pp. 152–162). Transaction Publishers.

Patino, B. 2019. *La civilización de la memoria de pez: Pequeño tratado sobre el mercado de la atención*. Alianza Editorial.

Rahwan, I. 2018. Society-in-the-loop: Programming the algorithmic social contract. *Ethics and Information Technology*, 20(5), 5–14. https://doi.org/10.1007/s10676-017-9430-8

Richardson, R., J.M. Schultz and K. Crawford. 2019. Dirty data, bad predictions: How civil rights violations impact police data, predictive policing systems, and justice. *New York University Law Review*, 94(15), 15–55.

Rousseau, J.J. 1895. *Discurso sobre la economía política*. Tecnos.

Rubel, A. and A. Clinton Castro. 2021. *Algorithms and Autonomy: The Ethics of Automated Decision Systems*. Cambrigde University Press.

Saunders, J., P. Hunt and J.S. Hollywood. 2016. Predictions put into practice: A quasi-experimental evaluation of Chicago's predictive policing pilot. *J. Exp. Criminol.*, 12, 347–371. https://doi.org/10.1007/s11292-016-9272-0

Solon, O. 2017. Ex Facebook president Sean Parker: Site made to exploit human 'vulnerability'. *The Guardian*, November 9, 2017. https://www.theguardian.com/technology/2017/nov/09/facebook-sean-parker-vulnerability-brain-psychology

Turing, A.M. 1974. *¿Puede pensar una máquina?* Departamento de Lógica y Filosofia de la Ciencia. Universidad de Valencia.

Turkle, S. 2019. *En defensa de la conversación. El poder de la conversación en la era digital*. p. 30, Ático de los libros.

We are Social and Hootsuite. 2021. *Digital 2021. Global Overview Report*. The latest insight into how people around the world use the internet, social media, mobile devices, and ecommerce. We are Social and Hootsuite. https://wearesocial.com/digital-2021

Williams, J. 2021. *Clics contra la humanidad. Libertad y resistencia en la era de la distracción tecnológica*. Gatopardo.

Wong, J.C. 2017. Former Facebook executive: Social media is ripping society apart. *The Guardian*. December, 12, 2017. https://www.theguardian.com/technology/2017/dec/11/facebook-former-executive-ripping-society-apart

Uhls, Y.T. and P.M. Greenfield. 2011. The Rise of Fame: An Historical Content Analysis. *Cyberpsychology: Journal of Psychosocial Research on Cyberspace*, 5(1), Article 1.

M. Zhou, F. Wang, T. Zimmerman, H. Yang, E. Haber, L. Gou. 2013. Computational discovery of personal traits from social multimedia. *In: 2013 IEEE International Conference on Multimedia and Expo Workshops (ICMEW)*, pp. 1–6, https://doi.org/10.1109/ICMEW.2013.6618398

Žižek, S. 2018. *El coraje de la desesperanza. Crónicas del año en que actuamos peligrosamente*. Anagrama.

Epilogue

Santiago Giraldo-Luque and Cristina Fernández-Rovira

Who will be the Next God?

Different studies conducted in the recent past have shown that we feel insecure when we do not have a smartphone with us (Yu and Sussman 2020). Other research conducted in a similar field has pointed out that young people literally do not want to stop using social media because "they are afraid of being nobody", displaying their dependence on social media (Giraldo-Luque and Fernández-Rovira 2020). On the other hand numerous journalistic reports denounce the fact that many of the same people, mainly young people, feel a high level of pressure emanating from social media networks that leads to a presence of symptoms of anxiety, significant levels of stress or depression (Cadena Ser 2021). In these reports, young people have stated, for example, that: "Social networks are the perfect platform to devour us, so that our backpack of complexes grows, and we only look at the pretty part".

Similarly, psychologists specialising in new technologies (Turkle 2019) have explained that young people have lost many basic social interaction skills because of their use of smartphones and social media. Empathy, conversation, problem solving or creativity to deal with boredom are some of the skills being lost by new generations who have grown up with technology. There are also numerous reports and research conducted that report disorders related to sleep loss and sleep disturbance (Quimis Ávila and Villavicencio Cobos 2020), and dozens of other psychological, physical, and social changes (Yu and Sussman 2020) that are caused by prolonged and excessive smartphone use.

'Our Minds can be Hijacked: the Tech Insiders who Fear a Smartphone Dystopia'. This was the title of an article published by Paul Lewis in The Guardian, which made visible what many former employees of large technology companies were beginning to think about the monster of mental and social control that they had helped to create while working in companies such as Google or Facebook (Lewis 2017). The identification of anxiety or fear produced by a disconnection from technology is a growing symptom that the hijacking of the human mind has effectively taken place.

For a lot of young people, technological contact is what determines their daily lives. All decisions and actions are taken in relation to it. Whether directly through a mechanic mechanism pre-designed for the user (such as a notification) or indirectly using technology as the only channel for social interaction, it is the concentration of attention on screens, and mainly on social media networks, that prefigures or preconditions the decisions and behaviour of individuals.

Social media has gone through three stages in their attempt to control users, and in each of them the domestication of will and the control of the human mind has been carried out at a greater level of depth. The mathematical logic applied to economic liberalism says that the more data and information one has about a product (or users, in this case), the greater the capacity for strategic decision-making. Hence, as social media platforms have accumulated more and more information about users, they have advanced their ability to read the minds of the users and to design their media to hijack or manipulate these minds. The anxiety generated by this disconnection and the declaration that it is impossible to stop using the networks, clearly demonstrates users' dependence on the systems of control and social and emotional prediction that social media networks have become.

In their first stage of their launch and social connection between the beginning of the millennium and 2007, social media platforms harnessed the power of social networking, people's interest in generating and distributing content created by themselves, and the capacity for intercommunication so much so that we all wanted to try. In this way, they revolutionised the way humanity communicated. Suddenly, we all wanted to be part of the social network because it embodied a message of freedom of expression and alternative universes of social communication. Without much thought and without measuring the consequences, just as had happened with the Web a decade earlier (which ended with a technological bubble that burst the first internet), we all wanted to have an account on Facebook, a platform that, in its beginnings, was not massively and intentionally dedicated to collecting all possible user data to sell to third parties.

The second stage began with the outbreak of the economic crisis of 2007/2008 and lasted until the middle of the second decade of the 21st century. The period of absolutist transformation towards control denotes an accumulation of technology companies in the hands of a few, and the first phase of control and prediction begins. The technological companies took advantage of the precarious situation of communication companies, as well as the media, to capture and co-opt their audiences. From one moment to the next, the media went from being the recipients of advertising revenue to having to pay for their content to be better positioned in search engines and social media. Based on the massive collection of user data, social networks began to classify users into advertising clusters to sell individualised and characterised segments of users to advertising companies (control); and to design consumption trends and practices on the platforms to anticipate certain purchases or user behaviour (prediction).

In this second stage, social networks capitalised on Amazon's predictive model and managed to increase their power and influence not only regarding purchasing decisions, but also decisions linked to citizens' rights, such as voting (Williams 2021). Crucial democratic rights became marketable goods or data. Tracking and projecting user behaviour, from the billions bytes of data collected, turned social media platforms and big technological companies into demigods. Their ability to anticipate decisions catapulted their power and influence, but at the same time began to reduce the autonomy of their users. Citizens began to be swayed by the decisions that machines made for them. Netflix's recommendation model is perhaps the clearest example of it, although it is possible that some of you may have decided on some of your travel destinations based on a viral Instagram photo.

The period of control and prediction in which we currently live is indeed a period to worry about as humanity. The third stage, which began in 2016 and coincides with Brexit and a few months later with the election of Donald Trump as president of the United States, can be called the consolidation of the matrix of will control. Once social networks, as well as Google and Amazon, have the users' data, they have the possibility of shaping people's will and inducing their behaviour. These digital platforms have control over the behavioural trajectories and sufficient knowledge of their users' decision-making systems (55% of humanity) thanks to years of training and experimentation, using notifications, advertising inputs, attention calls and social pressure.

Anxiety about the uncertain future, normally delineated and resolved by a notification, a suggestion, or an automatic recommendation in the everyday decision-making system, has become a process of collective trauma. The management of uncertainty, previously controlled by the machine's decision, becomes a vital problem for those who have not been educated in this necessarily human practice. When he or she discovers that they have to make a decision of their own, they go into crisis.

At the same time, the values associated with quick, and immediate success, have ended up being uncritically adopted by a large part of humanity. These are the values likened with fame which is apparently easily come and leads to extravagant standards of living, led by Jeff Bezos and Elon Musk in their private, space-conquering egocentrism, and which are then projected onto the absolute happiness and perfect fantasy world of social media.

The situation of mental dominance or reflective detachment in users is notorious. It is not uncommon to see in university classrooms that students do not know what to do when professors pose a problem that requires some level of creativity or complexity. They ask for instructions, an example and a model they can follow. Many of them are incapable of articulating an organised discourse to defend an argument that exceeds 280 characters. The nine-second attention span model (Patino 2019) of Tik-Tok and Instagram is increasingly predominant among young people and adolescents, and the control of their will is reflected in the violent, racist, and homophobic practices committed on the streets, which they record with their smartphones and then, without any possible rational explanation, share on social media (to receive the approval of the flock with hundreds of likes).

At this stage of consolidation of the matrix of will control, social media networks become the rulers of human behaviour. They are even venerated and they are identified as problematic and with myriad flaws. Yet we still refuse to abandon them. On the one hand, they have managed to imbue 55% of humanity with an empty, instantaneous, uniform and easily influenced identity (the rise of surgeries to imitate filters or the exemplary types of body aesthetics are a demonstration of this). The identity of the like and the follower. On the other hand, most internet users have been domesticated to assume the mantra that if something is not registered and shared on social media, it does not exist.

Among the giant technological companies, which are capable of predicting and shaping the future (it would only be enough for them to decide that something was a trendy topic to induce massive behaviour), the competition in this third period of complex and subtle domination is centred on the domination of will. In the role-play of domination, GAFAMs want to go from being demigods to being gods. At stake is the competition to be the world's new Omphalos, its navel.

Social media platforms have succeeded in holding minds hostage in the 21st century. In the face of insecurity, users cannot find autonomous, creative, or innovative answers. In the hijacking of the will, the critical spirit is dormant and controlled, anaesthetised. It sleeps peacefully fed by echo chambers built for the emotional wellbeing of the users.

The new society pre-designed by social networks does not, however, seem the most interesting. While some tycoons have the whim to travel into space, the concentration of vaccines in a few (rich) countries keeps the world on tenterhooks because of the continuous waves of the COVID-19 pandemic. The concentration of economic activity is the greatest in decades. The technological oligopolies deny any rules of classical liberal economics and act as absolutist corporations in price fixing, in the denial of free information for decision making and in tax evasion for which they use tax havens. The far right (as Trump did) uses social media to detonate every single value of representative democracy, as well as the pillars of welfare states which, curiously enough, have been the ones that have allowed technology companies to enjoy creative freedom.

Children also use social media to humiliate other children, harass their teachers and release videos commodifying their bodies and imitating the models they see, of course, on social media. Likewise, children are being cyber-attacked by exploitative networks, pornography, and bullying. And to add to all this, social media only outsources services (with exploited and poorly qualified workers) to filter content that, under behavioural policies designed by themselves, do not comply with their rules. Facebook, Instagram, and Twitter were accused of "silencing" Israel's violence against Palestine (Del Castillo 2021), just as Facebook blocked the voice of women building peace in Colombia (La Marea 2020).

The third stage of the control exercised by the platforms has normalised and, in a way, made the use of social media mandatory to ensure socialisation. Despite the obvious dangers, it is almost impossible to find young people who do not use them. The contradiction inherent in their use, in which citizens contribute daily to their own control and to the multiplication of the threats, confirms the collective hijacking of the intellect and communal creativity.

In consolidating the matrix of will control, social media networks have succeeded in emptying, as well as shaping, the minds of users. What will these new gods of will fill them with?

References

Cadena Ser. 2021. Social networks are the perfect platform to devour us, to make our backpack of complexes grow. *Cadena Ser.* July 30, 2021. https://cadenaser.com/programa/2021/07/30/hoy_por_hoy/1627621768_912228.html

del Castillo, C. 2021. Facebook, Instagram and Twitter, accused of "silencing" Israel's violence against Palestine. *ElDiario.es.*, May 18, 2021. https://www.eldiario.es/tecnologia/facebook-instagram-twitter-acusadas-silenciar-violencia-israel-palestina_1_7945010.html

Giraldo-Luque, S. and C. Fernández-Rovira. 2020. Social networks and digital consumption in young university students: Attention economy and communication oligopolies in the 21st century. *Profesional de la Información*, 29, 5, e290528. https://doi.org/10.3145/epi.2020.sep.28

La Marea. 2020. Facebook blocks the voice of women peacebuilders in Colombia. *La Marea*, December 9, 2019. https://www.lamarea.com/2020/12/09/facebook-bloquea-lo-voz-de-las-mujeres-que-construyen-la-paz-en-colombia/

Lewis, P. 2017. Our minds can be hijacked: The tech insiders who fear a smartphone dystopia. *The Guardian*, October 6, 2017. https://www.theguardian.com/technology/2017/oct/05/smartphone-addiction-silicon-valley-dystopia

Patino, B. 2019. *La civilización de la memoria de pez: Pequeño tratado sobre el mercado de la atención*. Alianza Editorial.

Quimis Ávila, M.M. and E.E. Villavicencio Cobos. 2020. Factors Involved in the Prevalence of Insomnia as a Sleep Disorder in Social Isolation. Bachelor Thesis. University of Guayaquil. http://repositorio.ug.edu.ec/handle/redug/53048

Turkle, S. 2019. *In Defence of Conversation. The Power of Conversation in the Digital Age*. Attic of Books.

Yu, S. and S. Sussman. 2020. Does smartphone addiction fall on a continuum of addictive behaviors? *Int. J. Environ. Res. Public Health*, 17, 422. https://doi.org/10.3390/ijerph17020422

Index